ADVANCE PRA
TAKING THE HEAT

"Bonnie Schneider's *Taking the Heat* details how climate change impacts the things we care about—our health, our children—and provides resources to help us build a more just, sustainable world."

—Gaurab Basu, MD, MPH, codirector of the Center for Health Equity
Education and Advocacy and instructor at Harvard Medical School

"Without presence of mind, how can we become better stewards of our bodies and planet? In *Taking the Heat*, Bonnie Schneider has crafted a powerful playbook that offers immediate solutions to these urgent questions."

—Amy Reyer, PhD, author of the blog *The Art of Living Slowly*

"Bonnie Schneider combines her extensive knowledge of meteorology and her clear, compelling writing style to navigate readers through the sciences so they can see for themselves the impacts of climate change on their own well-being."

—Drew Shindell, Nicholas Distinguished Professor of Earth Science
at Duke University

"This excellent book does a great job of helping us understand unfamiliar experiences like eco-anxiety and climate grief. It belongs on the shelf of every mental and physical health professional and community leader struggling with how to help people cope with our increasingly out-of-control weather and climate."

—Linda Buzzell, coeditor of *Ecotherapy: Healing with Nature in Mind*

"This is an important book for our times that marries meteorological understanding with health hazards associated with climate change. I'm pleased that Bonnie Schneider is bringing valuable advice to the public's attention."

—Janet Lewis, MD, founding member of the Climate Psychiatry Alliance

TAKING *the* HEAT

HOW CLIMATE CHANGE IS AFFECTING YOUR MIND, BODY, *and* SPIRIT *and* WHAT YOU CAN DO ABOUT IT

BONNIE SCHNEIDER

SIMON ELEMENT

NEW YORK   LONDON   TORONTO   SYDNEY   NEW DELHI

SIMON
ELEMENT

An Imprint of Simon & Schuster, Inc.
1230 Avenue of the Americas
New York, NY 10020

First Simon Element trade paperback edition January 2022

SIMON ELEMENT and colophon are trademarks of Simon & Schuster, Inc.

For information about special discounts for bulk purchases, please
contact Simon & Schuster Special Sales at 1-866-506-1949
or business@simonandschuster.com.

The Simon & Schuster Speakers Bureau can bring authors to your live event.
For more information or to book an event, contact the Simon & Schuster Speakers
Bureau at 1-866-248-3049 or visit our website at www.simonspeakers.com.

Manufactured in the United States of America

10 9 8 7 6 5 4 3 2 1

Library of Congress Cataloging-in-Publication Data
Names: Schneider, Bonnie, author.
Title: Taking the heat : how climate change is affecting your mind, body,
and spirit and what you can do about it / by Bonnie Schneider.
Description: First Tiller Press trade paperback edition. | New York : Tiller Press, 2021. |
Includes bibliographical references.
Identifiers: LCCN 2021023886 (print) | LCCN 2021023887 (ebook) |
ISBN 9781982166076 (paperback) | ISBN 9781982166083 (ebook)
Subjects: LCSH: Human beings—Effect of climate on. | Climatic changes—Psychological aspects.
| Climatic changes—Physiological effect. | Environmental psychology. |
Ecophysiology. | Self-care, Health.
Classification: LCC GF71 .S35 2021 (print) | LCC GF71 (ebook) | DDC 304.2/5—dc23
LC record available at https://lccn.loc.gov/2021023886
LC ebook record available at https://lccn.loc.gov/2021023887

ISBN 978-1-9821-6607-6
ISBN 978-1-9821-6608-3 (ebook)

CONTENTS

FOREWORD

As a family medicine professor and physician for over thirty years, my research and teaching has focused on primary care, cancer screening, and public policy. I can remember the exact moment when I personally realized that climate change was a threat to human health. That insight put me on a new course in life. In 2008, the United Nations' Intergovernmental Panel on Climate Change issued its fourth report, featuring an entire chapter on human health. To learn more about it, I attended a meeting with experts from around the country, including a climate scientist, public health experts, and authorities on communication. As a doctor who had taken an oath to protect people from harm, I knew then that climate change demanded my attention.

Several years later, I founded the Medical Society Consortium on Climate and Health. This organization includes members of twelve major medical organizations, among them the American Medical Association, the American Academy of Pediatrics, the American Congress of Obstetricians and Gynecologists, and the American College of Physicians. In assessing physicians' experiences with the health effects of climate change and their attitudes about and interest in engaging with the problem, our organization mobilizes public opinion to support ceasing the damage we inflict on our climate, protecting those who are vulnerable, and claiming a healthier future.

Climate change is one of the most important issues of our time and has major health and health care implications. Our national group now represents more than six hundred thousand clinical practitioners, including state-level groups (Clinicians for Climate Action), affiliated health organizations, and thousands of individual health advocates who devote their time to pushing for measures that will stop the degradation of our climate and protect the growing number of people who face health risks. The voices of America's medical so-cieties have the potential to help reframe the dialogue—putting human health and well-being front and center in the conversation. This important message needs to reach the public.

That's why, when Bonnie Schneider first reached out to me about her proposal for this book, I was intrigued at the prospect: a compelling description of health in the time of climate change.

And now, after reading the completed manuscript, my intrigue has turned to admiration. This beautifully written book helps to fill the gap in our aware-ness of the health hazards we face in the wake of climate change. Whether describing the anxiety experienced by young people who mourn the envi-ronmental damage to our world, the heat-related illnesses children suffer after being exposed in ways that their parents don't recognize, or the unexpected infections spread by mosquitos and ticks, Bonnie's writing and analysis are crystal clear and thoroughly informative.

Bonnie's ability to take dramatic human experiences and turn them into moving vignettes that inform us about the health dangers presented by climate change has produced a lasting resource that the lay public and health profes-sionals alike will find captivating. Psychiatrists caring for teenagers, internists treating elders, and public health professionals working to prevent heat illness will recognize their experiences in the stories they read here.

Throughout the insightful chapters of this book, readers will find both phy-sicians' and patients' perspectives on emerging climate-related health threats like the uptick in cases of Lyme disease and the increase in ER visits for those with lung and heart problems due to smoke from wildfires. Even better, this book provides positive prescriptions to help readers care for themselves and others.

It is my hope that *Taking the Heat* will not only educate more people about the harms of climate change but also inspire readers to seek the changes we need to move into a more modern era that benefits from cleaner, unpolluted air and water.

—Dr. Mona Sarfaty, MD, executive director of the Medical Society Consortium
on Climate and Health at the Center for Climate Change Communication,
George Mason University, Fairfax, Virginia
www.docsforclimate.org

INTRODUCTION

The human body has a natural way to manage heat. When our internal temperature rises from its normal set point of around 98.6 degrees Fahrenheit, an innate physiological cool-down process begins. Inside the brain, the almond-sized hypothalamus signals to blood vessels along the surface of the skin to dilate. Warm blood rushes to the body's extremities. Liquid heat is released in the form of sweat. As this moisture evaporates, our bodies cool.

But what happens when this instinctive process *doesn't* work?

If the body is unable to cool, its internal temperature continues to rise, potentially to extreme levels above 104 degrees. The skin becomes dry, turns red, and feels hot to the touch. As the heart pumps faster, dizziness or disorientation sets in. Speech comes out slurred. Within moments, a life-threatening situation has emerged.

This dire example is known as *heat stroke*, the most severe type of heat-related illness. It can be fatal, but even its milder version, *heat exhaustion*, can interfere with normal brain function if left untreated.

As global warming accelerates, dangerous scenarios like these may become more common. Extreme heat will impact more people and with greater frequency. Researchers at Climate Central project that average summer temperatures in most American cities are going be seven to twelve degrees hotter by the end of the century.[1]

Future grim heat predictions are only part of the story. The air we breathe, the water we drink, the food we eat, and the sanctity of our emotional well-being are all tied to the stability of our natural surroundings. Our environment can be healing or harmful.

Toxic wildfire smoke; supercharged allergy-inducing pollen; bacteria-contaminated water; disease-carrying insects; and devastating natural disasters have all been linked to climate change.

Doctors report the ensuing health consequences of global warming are already impacting their patients.[2] The problem is, most people are unaware of them. A recent survey showed that the majority of Americans have not yet considered how climate change may impact their health; only 32 percent could name a specific way in which it might affect them.[3]

This book will help you identify health hazards connected to climate change and provide you with expert advice on how to mitigate your own personal risk. But first, we have to fully understand what we're up against.

Taking the Heat

According to climate scientists, we are now in the warmest period in the history of modern civilization, and it's projected to get even hotter.[4] Scientists are in near-total agreement that over the span of more than a century, from 1880 to 2020, human activities have caused approximately 1.0°C (1.8°F) of global warming above preindustrial levels.[5] At this current rate, climate scientists project the Earth will continue to warm by 1.5°C between 2030 and 2052.[6]

This magnitude may be hard to understand because when we think about a mere one-degree Celsius rise (equivalent to an increase of about 1.8 degrees Fahrenheit) in daily or weekly weather, it might not seem like a big deal.

Viewed through the lens of climate change, though, that one-degree Celsius difference is significant.

Unlike *weather*, which is the day-to-day state of the atmosphere, *climate* comprises long-term averages and variations in temperature, measured over an extended period.

Throughout history, scientists believe the Earth's temperature fluctuated by no more than a few tenths of a degree (Celsius) from year to year. This means, in a geological context, that the recent global warming of 1°C in less than 150 years is an unusually large temperature change in a relatively short period of time.

Scientists also believe global warming is accelerating. The planet's seven warmest average annual temperatures have all occurred after 2014.[7]

The existing and forecasted global rises in temperature pose enormous environmental risks for your personal health and well-being. Some of these effects you may already be experiencing, while others have yet to be detected.

Even the healthiest among us are at risk. However, there are disparities in susceptibility. The elderly, the poor, children, and communities of color are disproportionately affected. This book also examines the inequity that exists for those who are most vulnerable.

Why I Wrote This Book

Throughout my career as a television meteorologist—whether I was forecasting severe thunderstorms on local TV stations or covering tornado outbreaks nationally on CNN—relaying important messages with urgency has been paramount. Extreme weather can pose an imminent threat to people's lives.

One of those dangerous events occurred late in the summer of 2019. I was on MSNBC, reporting live from the New York City studio on Hurricane Dorian, a category five storm bearing down on the Bahamas. Even if you didn't hear me describe how massive the hurricane was or the torrential downpours it produced, a quick glance at your TV screen confirmed the disaster: homes were destroyed, people were stranded or missing, families were torn apart.

Terrifying images on television, breaking-news banners, and bold weather graphics communicate the immediate danger of natural disasters. Relaying the same urgency when talking about climate change is more challenging. Rather than a dramatic and instant impact, climate change is a slow, insidious strain on our micro-environments.

Whether seasonal or ever present, this book will help you identify the harms you now face and what unseen hazards may lurk in the future. I've consulted leading experts, doctors, and scientists to canvass the latest research. I analyzed multitudes of data and studies, and spoke with individuals across the country so you could read and relate to their personal health stories. In the chapters ahead, you'll learn effective ways to mitigate risk—including expert strategies to lower your likelihood of contracting certain diseases.

As we now know, optimal wellness is achieved by taking care of not only the body but also the mind. That's why this book will also provide you with healing hacks for whole-person wellness—which includes prioritizing mental health.

Feelings of depression and anxiety have been linked to climate change concerns—especially in children and young people.[8] There's even a specific term for these fears: *eco-anxiety*. Parents will gain expert insight on how to discuss eco-anxiety with children, no matter what age those children are.

And all readers will discover the science-backed benefits of connecting spiritually with our outdoor environment as a form of self-care. Meditative walking, gardening, and planting can be calming and even restorative for those suffering from post-traumatic stress following a natural disaster. According to experts, people who live through a hurricane, for example, have an urgent need to reconnect and "trust" nature again as they seek relief from lingering emotional torment.

In fact, mindfulness and meditation have payoffs far beyond managing PTSD. For those with seasonal depression, stress-induced insomnia, or eco-anxiety, learning to be still and spend even a few minutes engaged in the present moment can be therapeutic. This book provides simple tips and advice to incorporate these practices into your everyday life.

It may seem counterintuitive to feel grateful at a time when things seem so uncertain. Yet the science on gratitude is astounding. Research suggests that when individuals practice gratitude, regardless of external circumstances, brain pathways strengthen at the neurotransmitter level.[9] When practiced regularly, gratitude can promote feelings of happiness and anxiety reduction. This important wellness key is highlighted throughout this book. You'll find advice on journaling and expressing gratefulness through meditation in the pages ahead.

My goal is for you to gain environmental awareness, to learn precautions, and, most important, to feel well—in mind, body, and spirit.

Bonnie Schneider

1.

ECO-ANXIETY

When the *Oxford English Dictionary* named *climate emergency* as its 2019 "Word of the Year," it was among a short list of other worthy contenders—including a relatively new term, *eco-anxiety*.

Defined as the fear of climate change and its impact on future life on this planet, eco-anxiety, especially for children and Generation Z (those ages eighteen to twenty-five), is a genuine source of mental anguish. In its milder forms, eco-anxiety might manifest as disappointment and frustration. But in the most severe cases, climate fears can overwhelm people to the point of unrelenting insomnia, difficulty maintaining daily functioning and self-destructive behaviors like substance abuse or self-harm.[1]

Parents are increasingly seeking advice on how to allay their children's fears about rising seas, melting ice caps, and destructive storms. In just the past few years, eco-anxiety has become a driving force in the global conversation about climate change—even prompting universities to create courses on how to handle "climate grief."

Mental health experts believe there are ways to cope with eco-anxiety. This starts with validating climate concerns, easing unmanageable levels of anxiety, and discovering new outlets for proactive endeavors to improve our planet's health.

Thoughts of Environmental Doom

"I watched this entire hillside of giant evergreens fall, one by one," twenty-year-old college student Barbara "Bee" Elliott said, recalling the barren site littered with overturned trees. "I felt as though it was ripping my own heart out from the soil as they fell because of our inconsiderate and dominating development."

The University of Washington Bothell undergraduate says her climate concerns escalated when she watched a land development project get underway near her apartment complex. As construction workers dug into the earth, laying concrete and asphalt, tall trees were knocked down in the process. Bee witnessed firsthand what she considered to be a "hurtful desecration of the natural environment."

As the project moved to its final phases, Bee's disappointment gave way to frustration. She felt no one was considering the needs of the trees. That's when she decided to do something. Bee wanted to alert people in the area as to what was happening all around them.

"I sawed off an inch of a tree stump still standing, sanded it down, and painted a crying eye being pulled open on the stump," Bee said. "This was to illustrate that these evergreens wanted us humans to wake up and open our eyes!"

Bee shared this experience the same year that one of the loudest and most influential voices for Generation Z, Greta Thunberg, spoke to world leaders at the United Nations Climate Action Summit in September 2019. The Swedish climate activist had previously revealed her feelings of eco-anxiety to the world: "When I was about eight years old, I first heard about something called 'climate change' or 'global warming.'"[2] Greta said she couldn't understand why adults weren't doing more to stop climate change.

Greta's message resonates with Gen Z. "I think Greta comes from a real place of concern that a lot of kids are feeling right now," twenty-three-year-old Maisy Roher said. The recent New York University graduate shared her own struggle with eco-anxiety in an interview with *Vice*. "I guess the despair started when I was eighteen, and I began learning about how much the Earth was changing, and I'd have full-blown panic attacks about the Arctic sea ice melting and the polar bears starving."[3]

Over the past few years, mental health professionals worldwide say they've seen an uptick of patients—from young adults to parents of small children— seeking help for eco-anxiety. "We're hearing more and more about this," said San Francisco–based psychiatrist Dr. Robin Cooper, MD, who's been in private practice for thirty-eight years. Dr. Cooper is a cofounder and steering committee member of the international Climate Psychiatry Alliance, a group of psychiatrists dedicated to educating the profession and the public about the mental health impacts of climate change.

Symptoms are often similar to clinical anxiety, Dr. Cooper said. "Eco-anxiety is a shorthand term for many extremely powerful and at times overwhelming feelings: fear, anxiety, distress, anger, sadness, and depression, which can merge into a sense of hopelessness."

The Mental Health Classification

Eco-anxiety, psychiatrists say, is not an "official diagnosis," meaning it's not classified as a mental illness or psychological disorder. The term is used to describe the focus of the patient's distress. Whether climate change is experienced indirectly or directly, it can translate into impaired mental health, resulting in depression and anxiety.[4]

In March 2018, a Yale University poll pointed to a rise in climate anxiety. Compared to a similar survey in 2014, nearly double the respondents (21 percent) reported being "very worried about the effects of global warming."[5] A more recent poll by the American Psychiatric Association (APA) found that young people were more likely to be concerned about climate change's impact on mental health than older adults: 67 percent of Gen Zers (eighteen to twenty-five years old) and 63 percent of Millennials (twenty-four to thirty-nine years old) responded that they were "somewhat or very concerned about the impact of climate change on their mental health," compared to 42 percent of Baby Boomers (fifty-six to seventy-four years old) and 58 percent of Gen Xers (forty to fifty-five years old).[6]

In a 2019 interview with *Medscape*, Dr. Lise Van Susteren said she would like the American Psychiatric Association (APA) to establish a subspecialty in climate psychiatry. This potential area of concentration would feature specialists who were trained in climate issues and could treat patients with eco-anxiety, and also offer targeted counseling to assist communities facing natural disasters.[7]

Pollution in Paradise

Twenty-one-year-old Katie Hearther grew up in a place many consider pure paradise: Hawaii. But even the most beautiful tropical destinations can be soiled by the effects of ocean pollution. Katie volunteered to help clean up the litter on Kahuku Beach with other teens while she was in high school. The experience was life changing.

Located on the Big Island's southern tip, Kahuku Beach was once a pristine spot of Oahu. Now, though, the shoreline is infamous for another reason: its massive accumulation of plastic trash and marine debris. As *Thrillist* reports, despite occasional sea turtle and monk seal sightings, "ocean currents and strong winds push plastic garbage and fishing industry waste onto the beach and into the rocky shelves."[8]

Katie described Kahuku as her personal ground zero. Up close, conditions at the beach were worse than she imagined. She painstakingly sifted sand with nets to remove tiny pieces of plastic. Each scoop was filled with so many particles that progress was slow and frustrating. Despite her time, focus, and energy, she only managed to clear a small area of the beach. She felt like her efforts didn't matter. "This snowballed into my every waking thought," Katie said. "There were terrible things happening to the ocean, and I felt powerless to stop it."

Katie, who also suffered from general anxiety, says she could not shake intrusive thoughts of the planet's perilous state.

"It can be extremely paralyzing," she said. "On the best days, it's like a quiet buzz in the background of my life. On the worst days, it is difficult to get out of bed because the weight of the global issue that is climate change is just so heavy. I can't articulate how every day is a battle to figure out how I can possibly be happy when no one cares that the world is on fire."

Katie says this fear influences her lifestyle choices and long-term thinking. She decided to dedicate her college studies to marine science, hoping to move things forward in a positive direction. She does everything she can to minimize her carbon footprint. But Katie feels these actions are not enough, as they will only have what she calls a "minuscule impact."

"I've already decided I'll never have children of my own," she confided. "I don't discuss these feelings often . . . but that's my reality."

That may sound drastic, but some recent studies point to climate change concerns influencing young people's choices in family planning. One study found that 38 percent of Americans between the ages of eighteen and twenty-nine believe climate change can factor in to a couple's decision to have children.[9] And a 2018 *New York Times* poll found that 33 percent of young adults say they're expecting to have fewer children than their ideal because they're worried about climate change.[10]

In an online interview with the Sierra Club, therapist Ann Davidman said that when the topic is explored further, additional underlying anxieties can be revealed. "It's easier and more socially acceptable to say 'climate' than 'I'm really ambivalent about having children.' . . . We often get judged and shamed for not knowing."[11]

Nightmares of "Climate Collapse"

Bee and Katie share several commonalities: they're both students at the University of Washington Bothell, and they each have friends and acquaintances who have categorized their views as alarmist or overdramatic. Katie said people often tended to be dismissive when she opened up and expresseed her thoughts about climate change. "There's a misconception that climate change is an issue of the future, instead of a global problem that's happening now," she said.

Toxic wildfire smoke, a future planet with no trees, and "total climate collapse" are fears that kept Bee Elliott up at night. Originally from Southern California, Bee said it was hard to express these feelings with friends, as some didn't take them seriously.

Katie and Bee found like minds in each other and support on campus. An innovative new college course offered an outlet for students like them to vent their climate concerns and help them discover new, productive ways to cope with their eco-anxiety.

Professor for Change

Natural disasters that would normally affect someone perhaps once in a lifetime impacted a University of Washington professor and her family *several* times in a matter of only a couple of years. For climate educator Dr. Jennifer Atkinson, watching this sequence of unfortunate extreme weather events unfold was transformational.

It started in 2016 when wildfire outbreaks engulfed the communities surrounding her mother's home outside of Paso Robles, California. Then another fire in October 2017 brought flames frighteningly close—within a quarter mile of the family home. Only a slight shift in the wind at the last moment spared the dwelling. The neighbors' homes were not so lucky.

This 2017 fire was part of one of the most destructive outbreaks that season, known in Northern California as "Wildfire Firestorm," a group of 250 wildfires that burned for weeks.[12]

Just a couple of months later, Dr. Atkinson returned to California for the holidays. She visited her family in Santa Barbara just as another massive fire was breaking out. Ripping across Ventura and Santa Barbara Counties, the Thomas Fire burned almost three hundred thousand acres before being fully contained, making it the largest wildfire in modern California history at the time.[13]

Dr. Atkinson recalled there was ash everywhere. Her young nieces and nephew wore face masks to go to school that December. "I remember walking out of *The Nutcracker* with them—this iconic symbol of the winter holidays—and my youngest nieces asked if they could make snow angels in the ashes on the ground," Dr. Atkinson recalled, wincing. "We ended up evacuating because smoke levels were so dangerous for the kids."

As if fires weren't enough to deal with, another type of devastating natural disaster began to unfold.

On the morning of January 10, 2018, flash-flooding triggered a massive mudslide along the burn-scarred slopes of the Santa Ynez Mountains. Witnesses said the avalanche of dirt, boulders, and debris sounded like a freight train as it crashed down the mountainside, taking with it everything in its path: trees, cars, entire homes, and people.[14]

"These massive mudslides that killed more than twenty people in Montecito," Dr. Atkinson said. "The hillside just melted away because all the vegetation holding it together was gone. . . . It was all disconcerting."

The impact of these consecutive natural disasters weighed heavily on Dr. Atkinson. She channeled these experiences into creating a college course to tackle the subject of climate change in a more personal way—focusing on environmental grief and eco-anxiety. Her goal was to explore the complex emotional and ethical issues of the climate crisis, as well as to offer a supportive space for students to discuss environmental injustice, to navigate despair, and to find hope in an ecological loss.

Not for "Snowflakes"

In 2017, when the media first reported on the existence of Dr. Atkinson's seminar, "Environmental Anxiety and Climate Grief: Building Resilience in the Age of Consequences," some of the initial feedback was not very complimentary.

"The comment section of the *Seattle Times* filled up with responses mocking our students as 'snowflakes,' 'wimps,' and 'coddled babies' who needed to 'grow up.' One person wrote, 'Do the students roll out nap mats and curl up in the fetal position with their blankies and pacifiers while listening to her lectures?'"[15] Dr. Atkinson said.

That initial skepticism was short-lived. At the University of Washington and within the community overall, the course has been positively received. "I got an outpouring of letters from people saying, 'Thank you; just knowing that this seminar exists and that there's a name for this condition makes me feel less abnormal. I'm not alone,'" Dr. Atkinson said.

Registration has swelled since the course was initially introduced, and it has even converted some former climate-change naysayers. "It's been extremely popular, with every seat filling up. This quarter I have students from all three UW campuses [Seattle, Tacoma, and Bothell] enrolled and making the trek up to our campus each week to take part."

Katie, who enrolled in the class in the spring of 2018, said she doesn't hold back her feelings in this class. "I can simply embrace my grief in a safe environment."

The students read works of literature and poetry and view art to "humanize the issue so it's not just a statistical problem, it's not just something we represent with charts and graphs. And that helps connect this issue to our own lives and experiences," Dr. Atkinson explained.

Learning about environmental and social justice movements provides a road map for moving from grief to action. The class examines individuals and communities that have overcome setbacks and loss and have developed resilience to stay engaged in social movements over the long run. Dr. Atkinson believes civic action can be a therapeutic antidote to climate grief.

This sentiment has gained traction internationally. Young people gathering to have their voices heard about climate change has been part of a global effort driven in part by eco-anxiety. The Global Climate Strike, held from September 20 to 27, 2019, saw a record 7.6 million young people march in protest and demand action from world leaders.[16] Many schools permitted absences for students so that they could participate.

Bee said participating in climate-action endeavors makes her feel less depressed about climate change's potential future impacts. "This gives me so much hope to meet other people who care, and who are taking this as seriously as I am."

Group Support

While Bee and Katie found a like-minded community at their university, older millennials in their thirties and Generation Xers who are now in their forties and fifties discovered other ways to express and address their eco-anxiety. Forty-five-year-old Risa Robertson, for example, found support through social media.

Active on Instagram, the California resident captions her colorful nature photographs of Lake Sabrina in the eastern Sierras and Yosemite National Park with hashtags #ecoanxiety and #getoutside.

On Instagram, #ecoanxiety is a prominent hashtag, with about ten thousand posts. Risa said through following that hashtag in particular she's connected with people as far away as Australia who share her concerns about drought and bushfires.

From fire tornadoes on South Australia's Kangaroo Island to red skies and barely breathable Sydney air, images and videos have brought new urgency to the issue and provided an outlet for sharing climate grief and fear.

Psychiatrist Dr. Cooper says that, whether through the looser format of engaging with others on social media or more structured group settings, these connections are a crucial tool to cope with eco-anxiety. "Finding a group that fits one's style, temperament, skills, talents, and interests can sustain the engagement. There are so many."

The Good Grief Network is one of them. A nonprofit network of nationwide support groups offers a ten-step program that helps those dealing with climate grief. Like the framework of Alcoholics Anonymous, each step delivers a specific concept for how to cope.

Good Grief Network's founders LaUra Schmidt and her wife, Aimee Lewis-Rowe, established the group in Utah back in 2016 and now host hundreds of meetings and training sessions, and offer online courses through their website, goodgriefnetwork.org. They facilitate the metabolization of heavy feelings, which can put us at risk for burnout and cause us to fall into despair, eco-anxiety, or depression.[17]

Leah Hogsten was an early participant of the Good Grief Network group support when it began a few years ago under the backdrop of Utah's Wasatch Mountains. She told Yale Climate Connections that she appreciated the opportunity to talk with like-minded people—and discuss solutions.[18] "You've vented and gotten some worries off your chest, and now you have a better understanding of what you can do as an individual," Leah said.

Talking to Young Children about Eco-Anxiety

In the popular HBO TV series *Big Little Lies*, based on the bestselling book by Australian author Liane Moriarty, an episode revolved around a child's fears about climate change. Amabella, the eight-year-old daughter of Renata, has a panic attack after learning in school that "the world is doomed." Renata, displeased with the dire way the subject was brought up in the classroom, seeks to allay her daughter's eco-anxiety.

How to talk to kids about this topic is a growing concern for parents in real life too. In the 2020 article "Climate Anxiety in Young People: a Call to Action," published in the *Lancet* journal, researchers underscored the urgency to address children's climate concerns: "Young people are agents of change, our future leaders, and most likely to succeed in improving planetary health. Thus, making investments to improve their mental health and well-being will provide dividends now and in the future."[19]

How Should Parents Talk to Kids About Climate Change?

Start with the facts, according to the experts at NASA's Climate Kids website, https://climatekids.nasa.gov. The site is a helpful tool for parents, offering ways to approach the subject in kid-friendly language, along with videos and games children can play to learn more.

Experts advise being especially careful about using "doom and gloom" language with very young children. Antioch University's David Sobel, an environmental educator and faculty member, explained in an interview with *HuffPost* that he likes to use the expression "no tragedies until fourth grade."[20]

According to Sobel, calamities encompassing significant Earth-scale problems can be harrowing for younger kids to process. Topics like rainforest destruction and ozone depletion may be too complex for little ones. He explained that children's minds experience a developmental shift around the fourth grade, broadening their understanding and awareness. This is when many kids begin to develop the capacity to think rationally and use reason over emotion.[21]

That thought process may lead the child to question: *What can I do about it?*

Therapists advise parents to explain to kids what they can do to help make a difference.

"Children of various ages seek ways their own actions can make an impact. Recycling, participating in environmental cleanups, and even planting a tree are activities families can do together," according to the Rainforest Alliance, an international nongovernmental organization (rainforest-alliance.org).

The Sierra Club (sierraclub.org) offers this list of youth-centered climate groups for children and parents:[22]

1. Sunrise Movement: https://www.sunrisemovement.org
2. Zero Hour: http://thisiszerohour.org
3. Power Shift: https://www.powershift.org/members
4. Sierra Student Coalition: https://www.sierraclub.org/youth
5. Youth v. Gov: https://www.youthvgov.org
6. Earth Guardians: http://www.earthguardians.org
7. Mothers Out Front: https://www.mothersoutfront.org
8. Climate Parents: https://www.sierraclub.org/climate-parents

For kids that are old enough to use social media, environmental platforms can serve as a means to spread awareness about the climate crisis. Some helpful Instagram accounts to follow are @youthgov, @nextgenamerica, and @climateoptimist.

Nature Immersion for Kids

Outdoor activities or just spending more time outside to connect with natural surroundings is also recommended to cope with eco-anxiety. Eugene, Oregon–based therapist Dr. Patricia H. Hasbach told *Sierra* that nature immersion "offers a deep, sincere connection to that part of themselves, recognizing that they *are* the natural world."[23]

Through these experiences, experts say, children develop resilience that will help with their current eco-anxiety and keep them engaged in climate change effects when they grow up.

Researchers at Cornell University found that when children younger than age eleven spend time in nature—hiking, camping, hunting, or fishing, for example—they become adults who care more about the environment than those who didn't have that early exposure.[24]

Eco-Therapy

Connecting with nature to cope with climate concerns isn't just for kids. It's beneficial for adults as well. One version of this type of stress management is called eco-therapy. Eco-therapy is the modern equivalent of ancient customs focused on appreciating and honoring the Earth, and lives at the crossroads of ecology, environmental activism, and psychology.[25]

In Santa Barbara and Los Angeles, California, eco-therapist Linda Buzzell helps her patients connect their minds to nature in order to manage anxiety. She also teaches other therapists how to implement eco-therapy in their own practices.

"In my ecopsychology and eco-therapy teaching work, on campus and online, nature connection practices can be extremely beneficial in helping us cope with the escalating bad news of our planet's health," she reports.

Goodtherapy.org, a website that can help you locate mental health counselors in your area and gather related information, breaks down the various methods of eco-therapy.[26]

Some doctors described the mental health benefits of eco-therapy as a "nature pill." There's science to back it up. In a study published in *Frontiers of Psychology*, researchers from the University of Michigan examined cortisol levels (the stress hormone) of participants for two months. The subjects spent at least ten minutes outside in nature, three times a week.[27]

"We know that spending time in nature reduces stress, but until now, it was unclear how much is enough? How often should we do it? Or even what kind of natural experience will benefit us?" said Dr. MaryCarol Hunter, an associate professor at the University of Michigan who was the lead author of this research.[28] "Our study shows that for the greatest payoff, in terms of efficiently lowering levels of the stress hormone cortisol, you should spend twenty to thirty minutes sitting or walking in a place that provides you with a sense of nature."

Eco-Therapy Practices

- **Nature meditation:** Identify something in nature, and then spend a few minutes contemplating why it attracted you and how you relate to it.

- **Horticultural therapy:** Plants and garden-related activities can be used to promote well-being. These include digging soil, planting seedlings, weeding garden beds, and trimming leaves.

- **Animal-assisted therapy:** Time spent with animals like horses or dogs can help the healing process.

- **Physical exercise in a natural environment:** Walking, jogging, cycling, or doing yoga in a park foster increased awareness of the natural world. Recommended for reducing stress, anxiety, depression, and anger.

- **Conservation activities:** Restoring or conserving the natural environment can provide purpose and a sense of hopefulness, particularly when practiced in a group setting. This may help foster a sense of belonging and connectedness with others.

Source: GoodTherapy.org

Walking Mindfully

Berkeley, California–based eco-therapist Ariana Candell follows these principles in her thriving practice. "I will take individual clients outdoors into the hills, into the parks, and we'll do our sessions out there. Instead of it just being both of us in the room, we will actually be among the trees, the ground, the pine cones, the birds," she said, smiling.

"There's a relationship between us and nature that sometimes we don't realize and have enough gratitude for, or take the time to cultivate."

As the founder of the Earthbody Institute, Ariana Candell offers guidance (below) on how to connect with nature in order to reduce anxiety while taking a walk outside.

Professional Therapy

In terms of self-care techniques, psychiatrist Dr. Robin Cooper, assistant professor in the department of psychiatry at the University of California San Francisco, said meditation, exercise, and deep breathing have helped some of her patients.

"However, for more significant difficulties that impair functioning, therapy is indicated," Dr. Cooper said. "Therapy can be useful in breaking down the sense of hopelessness through the essential experience of being listened to, having concerns validated, managing paralyzing symptoms, and having help in challenging destructive cognitions and mobilizing for effective behaviors."

Katie Hearther credits professional therapy as an essential part of her overall strategy to cope with eco-anxiety. She says she continues to find techniques to help manage her feelings. "I quickly found that my existential dread in the face of climate change is not unique. I am trying hard to work on living in the moment, being thankful for what I have, and regularly attending therapy," Katie said, summing up a three-tier strategy that's been helpful for her.

Bee said that becoming more involved in pro-environment causes has allowed her current thoughts to trend toward hopefulness and away from fear. "I'm not letting myself be scared and turned away by the crisis. Instead, I'm embracing this unprecedented time and accepting it as an invitation to a more meaningful life with more purpose."

Walking Mindfully in Nature Tips

- **Feel the shift:** from the concrete into a little bit more of a natural setting and make that awareness in your mind so you can feel a shift in your body and your mind that says "You're going to be in a more relaxing state of being now."

- **Observe nature:** Be as present as possible to what you are noticing. It could be in the blossoms; it could be in the new green leaves, the scent of the air, the wind, or the reflection of the sunlight.

- **Don't rush:** Take your time, so you're not just running past and rushing around. You're slowing your mind and body down. Shift gears as you get away from what you were doing at home.

- **Pay attention:** Feel and hear the sound of your feet as they hit the ground. Walk with awareness versus staying in your head and thinking of all your problems.

- **Be open:** to what's happening around you. To be truly healthy and whole, we need to live with the conscious awareness that we are connected with all of life: the waters, lands, air, and creatures.

(Source: Author interview with
Ariana Candell of the Earthbody Institute)

2.

NATURAL DISASTERS, TRAUMA, AND GRATITUDE

"For me, Sandy is never over," says Hurricane Sandy survivor Beth Henry of Long Island, New York. Even years later, safe in a new residence, the family experiences post-traumatic stress following their harrowing experience surviving the 2012 superstorm. "My kids still get scared every time there's talk of another hurricane in the forecast," she said.

The recurrent fears Beth and her family experience are not anomalies. Living through an extreme weather event like a hurricane can bring psychological impacts that build over time, lingering even years after the disaster. For example, in New Orleans, doctors are still currently treating the emotional anguish caused by Hurricane Katrina, which hit Louisiana back in 2005.[1]

In the actual moments during a traumatic or unexpected dangerous event, people's first reaction is usually a combination of shock and fear. During these instances, the body's "fight or flight" response is activated, unleashing a jolt of adrenaline and stress hormones like cortisol. This may cause someone to sweat and their heart to beat faster when faced with an emergency.[2]

That's often immediately followed by emotions like disbelief, confusion, and denial. According to the mental health experts, these are protective and adaptive reactions, part of *post-traumatic stress* (PTS).[3] In an interview with Health.com, neuropsychologist and teaching faculty member at Columbia University Dr. Sanam Hafeez, PsyD, explains that "PTS is considered a normal reaction to stress."[4]

As time passes, these residual feelings of anguish can vary for each person. For example, those who were evacuated or whose homes were completely destroyed may face more significant hurdles than those who encountered less permanent effects.

PTS sufferers may feel grief-stricken and despondent. They might find it difficult to concentrate, make decisions, or get a good night's sleep. Vivid memories may instill ongoing fear that the event is going to happen again. This can manifest in a heightened anxious state where the person is easily startled. Experts say PTS is more short-lived, though, in comparision to *post-traumatic stress disorder* (PTSD), which can be ongoing and last for years.

In these instances, professional counseling can be a source of healing. Resilience, the process of adapting well in the face of adversity, is built through taking proactive steps in storm preparedness, getting back to routines, mindful meditation, spending time in nature, and even practicing gratitude for what remains.

PTSD

PTSD is a diagnosable condition listed in the American Psychiatric Association's *Diagnostic and Statistical Manual of Mental Disorders* (DSM). It is essentially an anxiety disorder.[5] Not all who experience a traumatic event will develop PTSD. Many people will have a traumatic event in their lifetime (70 percent of the general population), yet only 7.8 percent develop PTSD in the aftermath.[6]

Researchers found that hurricanes led to higher rates and longer-lasting instances of PTSD, along with stress, anxiety, and depression, than other types of flooding events.[7] One study, conducted fifteen months after Hurricane Katrina, found that over half (52 percent) of subjects continued to experience poor mental health following the storm.[8] In the shorter term, at around six months after Katrina, survivors expressed symptoms consistent with a diagnosis of PTSD at a rate of 19.2 percent.[9]

This ongoing effect was also evident following Hurricane Andrew, another catastrophic hurricane that hit the US, in 1992. A study of impacted Florida residents found the prevalence of PTSD increased by almost 30 percent up to thirty months after the disaster.[10] Many individuals develop PTSD symptoms within three months of the trauma, but symptoms may appear later and often persist for months and sometimes years.

This gets to the heart of the difference between PTS and PTSD. The symptoms are similar, but the distinction lies in how long those symptoms last, and how intense they are.[11] For a person to be diagnosed with PTSD, symptoms must last for more than a month and cause significant distress or problems in the individual's daily functioning. PTSD often occurs with other related conditions, such as depression, substance use, memory problems, and other physical and mental health problems.[12]

Any person, even those who are psychologically healthy, may develop PTSD when exposed to a highly traumatic event.[13] While PTSD can occur in all people of any ethnicity, nationality, culture, and age, women are twice as likely as men to have PTSD.[14] Also children, adolescents, disaster volunteers, and individuals with prior trauma or pre-existing psychiatric disorders have been identified as special populations at risk for PTSD.[15]

Perceived Stress on the Mind and Body

How we perceive stress plays an essential role in how our bodies respond mentally and physically to it. Health scientists say we rely on our ability to use past experiences and information about our current environment to predict what might occur in the future. Our perpetual goal is to increase the odds of desired outcomes, while avoiding or bracing ourselves for potential adversity.[16]

Fear in itself can be protective. The human brain can recall past events, note present surroundings, and use this knowledge to assess how likely something is to happen and when. Uncertainty is only diminished to the degree to which we believe we can prepare in our minds for what's ahead. Often, we don't know for sure what's coming, and that doubt can lead to worry. Thoughts spiral, tilting us negatively toward a "worst-case scenario" mindset. This can result in ongoing anxiety.[17]

For those with PTSD, just recalling details of the traumatic incident can trigger physiological responses in the body, such as the release of inflammatory hormones into the nervous, endocrine, and immune systems, causing a stress overload.[18]

According to the Post Traumatic Stress Alliance, an association of advocacy and professional organizations for individuals suffering from post-traumatic stress disorder,[19] PTSD sufferers can experience symptoms in both the mind and the body. Emotions of anger, depression, irritability, or sadness can be accompanied by fatigue, increased perspiration, high or low blood pressure, and trouble digesting food.[20]

The brain's *amygdala*, where emotions and fear are processed, may become hyperactive when impacted by PTSD.[21] For example, if you live in an area that's repeatedly beset by wildfires or hurricanes, seeing nearby communities evacuate each year can trigger distress. For natural disaster survivors diagnosed with PTSD, these exposures can lead to chronic stress, heightened fear, and increased irritation.

Climate Change and Weather Disasters

Extreme weather events themselves are rare. The odds of having your home destroyed in an epic flood, being forced to evacuate due to a monster storm, or barely surviving a tornado outbreak vary depending on where you live. Some places, though, tend to impose higher risk through geographic location. Compared to all other US states, Alabama, Georgia, and Mississippi boast the dubious distinction of having the highest likelihood of experiencing a natural disaster, according to the CNN.com article "Which Natural Disaster Will Likely Destroy Your Home?" which cited research by the real estate information company RealtyTrac based on US Geological Survey and National Oceanic and Atmospheric Administration (NOAA) data on past storm incidence from more than three thousand counties in the US. Their research pointed to a greater probability of your enduring some type of natural disaster if you live in any of those three southeastern states.[22]

For those who reside in other parts of the country, however, the report wasn't exactly reassuring. RealtyTrac found that 55 percent of homes in the US are located in "very high" or "high" risk disaster zones.[23]

When you look back at the disastrous year of 2020 and evaluate the number of extreme weather events that impacted the country, the increased likelihood of your living in a "high risk" area for a natural disaster doesn't seem farfetched. This unforgettable year was marked by tragic losses due to the COVID-19 pandemic. It was also a time when a record number of weather disasters affected millions of Americans, according to the NOAA.[24]

For the past six consecutive years, weather events—like hurricanes, wildfires, tornadoes, and floods—have been especially devastating and costly, totaling at least ten billion dollars or more annually in the United States.

Climate scientists anticipate more extreme weather events going forward due to global warming. According to the Intergovernmental Panel on Climate Change, the current rate of global warming is projected to rise by 0.2°C per decade.[25] Their trajectory points to more intense storms occurring over the twenty-first century.[26]

Financially, the cost for reconstruction and repair following each natural disaster can be estimated and tallied. Emotionally, though, the price goes unchecked. In the aftermath of a storm, these quietly looming mental-health impacts can build over time.

The Wrath of Maria

In September 2017, the last thing Puerto Rico needed was *another* storm. Residents were in the midst of recovering from one just a few weeks prior that had left over sixty thousand residents without power.[27] Unfortunately, this next hurricane proved to be much worse.

Hurricane Maria struck Puerto Rico as a Category 4 storm with maximum sustained winds estimated between 130 to 156 mph. According to the NOAA, "meteorologists have no land-based records of Maria's precise maximum winds because the storm damaged most of Puerto Rico's wind sensors and a weather radar tower."[28]

The fierce gusts uprooted trees, stripping their branches bare of leaves as the storm tore through the island. Communication was utterly cut off. Ninety-five percent of the island's cell phone towers were severely damaged.[29] According to NOAA reports, Puerto Rico's aging electrical grid, already compromised by the previous hurricane, was destroyed. At the height of the storm, the entire island was without power.[30]

There were few accessible escapes. Twenty inches of rain caused massive flooding, with some spots receiving even higher amounts. Roads, traffic lights, and street signs were washed away. In Toa Baja, part of the San Juan metropolitan area, families awaited rescue on their rooftops.[31]

Recovery was excruciatingly slow. Only about 8 percent of roads were open a month after the hurricane. Five months later, a quarter of the island's residents still lacked electricity.[32]

Puerto Rico's nearly seven hundred thousand children faced death, loss, and economic hardship to varying degrees.[33] Less than a year after Maria, a study published in *JAMA Network Open* noted that more than 7 percent of children on the island met the clinical standards for PTSD, about twice the rate seen in the general population.[34]

To better ascertain the impact Maria had on its youngest victims, researchers surveyed around ninety-six thousand students in Puerto Rican schools. They found:

- More than 57 percent of children had a friend or family member leave Puerto Rico due to the storm[35]
- 45 percent reported damage to their homes
- 32 percent experienced shortages of water or food
- About 33 percent felt their lives were at risk during and after the storm

The study's authors note their findings may have even underestimated the extent of the problem.[36]

"These alarmingly high numbers signal that nearly every child in Puerto Rico was exposed to disaster-related risk factors during Maria," says psychologist Rosaura Orengo-Aguayo, PhD, lead author of the study and assistant professor at the Medical University of South Carolina's (MUSC) department of psychiatry and behavioral sciences. "We also know that these factors are some of the predictors of developing post-traumatic stress, depression, and anxiety down the line."[37]

Children's Feelings

The youngest survivors of natural disasters are often affected the worst emotionally. Unfortunately, children are more likely to have continued trauma-related symptoms.[38] Family routine disruptions, separation from caregivers due to evacuations or displacement, and parental stress after a disaster all contribute to youngsters' distress.[39]

Researchers studying the deadly 2011 tornado outbreak in Joplin, Missouri, found an increase in post-traumatic stress disorder rates, even two years later.[40] Young children (ages four to ten) struggled more than their older peers, according to their parents, particularly if the parents themselves were experiencing PTSD.[41]

Mental health experts advise parents to contact a professional if children exhibit significant changes in behavior or any of the following symptoms over an extended period (more than six months) after experiencing the trauma of a natural disaster:

1. **Preschoolers**—thumb-sucking, bedwetting, clinging to parents, sleep disturbances, loss of appetite, fear of the dark, regression in behavior, or withdrawal from friends and routines
2. **Elementary school children**—irritability, aggressiveness, clinginess, nightmares, school avoidance, poor concentration, or withdrawal from activities and friends
3. **Adolescents**—sleeping and eating disturbances, agitation, increased conflicts, physical complaints, delinquent behavior, or poor concentration[42]

Other symptoms not listed above include reexperiencing the disaster during play or dreams, anticipating that the disaster is happening again, general numbness to emotional topics, and increased startle reactions.[43]

Displacement

Climate change raises the potential for more people to face mental health impacts brought on by natural disasters.[44] This is particularly true for those who are displaced from their residences.

The home is a place that provides safety and security. When you're forced to leave, especially with little notice, taking with you only whatever you can carry, it can be devastating. Rising sea levels and increased instances of flooding, fires, and mudslides are projected to displace about 200 million people globally by 2050.[45]

Loss of place is not a trivial experience.[46] According to the 2017 report "Mental Health and Our Changing Climate: Impacts, Implications, and Guidance," those who are strongly connected to their local communities report greater happiness, life satisfaction, and optimism.[47] When those connections are severed, post-traumatic stress or its more severe and persistent form, PTSD, may result.

Solastalgia

Mental anguish can be experienced by people who don't lose their own homes entirely or even sustain any personal property damage when a natural disaster strikes. In fact, their residence may be untouched—but when repeated disasters happen to the same community, it can trigger mental anguish.

Imagine annual nor'easters with whipping winds that erode coastal beaches until there's no commercial waterfront left. Envision multiple subdivisions flattened by tornadoes each season, altering a suburban community block by block. Picture sequential wildfires scorching the same mountainsides, each time leaving less livable area behind.

Witnessing such damage to your surroundings, even indirectly, can elicit feelings of loss, grief, and helplessness. Psychologists describe these sentiments as *solastalgia*—a longing for a home community that has been reshaped by climate change to go back to the way it was before.

Solastalgia is a relatively new concept, and researchers emphasize that these feelings may not always occur immediately following a disaster. Solastalgia may come on more gradually due to the slow onset of changes that can reshape a person's environment. It's characterized by a sense of desolation and loss similar to that experienced by those who've been forced to migrate due to unforeseen circumstances.[48]

In an interview with the *Guardian*, Ashlee Cunsolo, the interim dean of the School of Arctic and Subarctic Studies at Memorial University of Newfoundland in St. John's, pointed out that "Indigenous people are particularly vulnerable to feelings of solastalgia due to their deep connections to their homelands and their practical daily knowledge of the local area."[49]

Children, Poverty, and Natural Disasters

Massachusetts-based pediatrician Dr. Aaron Bernstein says children who live in poverty are at greater risk for post-traumatic stress following natural disasters. Interim director of the Center for Climate and the Global Environment at the Harvard T. H. Chan School of Public Health, Dr. Bernstein has vast experience as a pediatrician treating children with various ailments, including trauma.

"One of the biggest risk factors for children facing ill health in the world, and in the United States, is poverty," he says, pointing to income disparities. "For kids who are living through disastrous weather events . . . as it is with so many other things . . . if you're rich and you have home insurance, and if your house gets destroyed, your parents really can say, 'This is going to be okay. We'll figure this out.' But if you're poor and you don't have insurance, and you're living in a community that has no economic activity, that adds another layer of stress and uncertainty."

Every year, 175 million children globally are affected by natural disasters, including floods, cyclones, droughts, heat waves, severe storms, and earthquakes.[50]

Media Influence

Turn on your TV, scroll through social media, read online news—when a natural disaster strikes, the impacts are continuously reported for weeks afterward.

Health experts say media messaging and public communications about extreme weather events and climate change can affect mental health and well-being.[51]

This is especially true for children. Besides what they see and read themselves, kids overhear adult discussions about tragic news stories. Mental health experts emphasize that the way parents and caregivers react to and cope with a disaster can affect how their children respond.

Young people who view media coverage may display symptoms of being afraid, worried, or anxious; they may also experience sleep disturbances or an inability to stop thinking about what they've seen or heard.[52]

"One of the things critical for children to do is to turn off the media following a natural disaster," said Dr. Joy Osofsky, PhD, a clinical and developmental psychologist and psychoanalyst and professor of pediatrics and psychiatry at Louisiana State University Health Sciences Center. "Children don't need to see these pictures or videos over and over again because that can be upsetting and counterproductive."

If parents allow children to watch television or use the internet, where they might come across such images, mental health experts advise parents to encourage communication and provide explanations. This may also include monitoring and appropriately limiting their own exposure to anxiety-provoking information.[53]

"It's important, depending on their ages," said Dr. Osofsky, "to ask them what they know after there's been a big storm. Ask them, 'What have you heard about it?'" This may help children feel more in control of their reactions.

School Support

Schools play a significant role in disaster recovery and are often the backbone of relief operations. They are where children can regain a sense of normalcy in their lives and receive psychological support while doing so.

Additionally, services can be delivered in schools without the stigma commonly associated with mental health interventions. Parents and families know and generally trust school personnel and processes.[54] Schools are typically one of the first organizations to resume operations after a disaster.

Dr. Joy Osofsky was one of the leaders of a study conducted by the LSU

Health Sciences Center's department of psychiatry that evaluated children's mental health one year after Hurricane Katrina. She saw firsthand how crucial school support is when it comes to helping children return to their routines.

In the aftermath of Hurricane Katrina, LSU researchers evaluated children who were evacuated to cruise ships just after the storm, those who returned to St. Bernard and Orleans parishes, as well as children who remained displaced. Their objective was to measure their levels of distress and need for psychological services.

"These children experienced a difficult evacuation and significant personal losses," said Dr. Joy Osofsky.

In Saint Bernard Parish, right outside of New Orleans, these distressed children went back to school months after Hurricane Katrina, one of the deadliest storms in our nation's history. Dr. Osofsky documented what happened on the first day of their return:

> When the levee broke in the Lower Ninth Ward, there was a lot of flooding. . . . Most schools were destroyed, but the top level of one of the three schools remained intact. So they set that up for students to return to school there in November of 2005.
>
> Our mental health team had been helping them with health support. We said to them, "We'll just be there when you reopen, not to get in the way." And so, very quietly, we stood by as people came into the school.
>
> After about fifteen or twenty minutes in, without any prompting, people were sharing their stories—it wasn't just the children, it was the parents, too. Despite the tragedy, they were so happy—not only to come back but just to see each other. To be together . . . The school was especially important for supporting the community and certainly for supporting children.

Data collected from children returning to St. Bernard and Orleans Parishes in the study showed that over 31 percent reported clinically significant symptoms indicative of depression and PTSD. The team found that of the displaced and returning children, 54 percent experienced symptoms that put them in need of further mental health care.[55]

Building Resilience

Resilience refers to a dynamic process of positive adaptation within the context of significant adversity.[56] It's considered a key element in lifting oneself up and enhancing one's quality of life.[57] In adults, resilience generally develops over time. But young people with less life experience can be taught how to tap into these coping skills.

How can we instill resilience in children? Experts suggest that everyday challenges present opportunities. These are the instances where kids may try and fail at something, then try again. Pediatrician Aaron Bernstein suggests that children be allowed to take some risks, "so they learn to be less afraid of threats and change. It's teaching them that they can figure things out."

He clarifies, "You don't want the child jumping off cliffs, necessarily, but healthy risk-taking involves a real challenge. These can present an opportunity for growth and push on the child's threshold of fears or concerns. That's good in general for any stress that comes your way. But it's imperative in the context of climate-related threats.

"There's a difference between when disappointment happens, and it's sort of out of your control," Dr. Bernstein said.

Resilience isn't just the ability to deal effectively with negative situations and to recover quickly; it's also mental preparedness for future problems and vulnerabilities.[58] That's when practicing risk-taking helps. It's the kind of thing where you're saying to the child when learning something new, "We're going to do this, and have confidence in your abilities," Dr. Bernstein said. For example, the first time a child tries to ride a bike may be tricky or even discouraging after some false starts. But succeeding—especially after those setbacks—is rewarding. And it was a risk worth taking.

Children also learn that sometimes disappointments are out of their control—even minor ones like a friend who cancels a playdate at the last minute. Dr. Bernstein explained that as children bounce back from these seemingly small setbacks, especially when they're unexpected, their inner resilience strengthens a little more each time.

Some evidence suggests that resilience is more abundant in people who have repeatedly been impacted by previous challenges, according to Dr.

Jonathan Purtle, policy dissemination and implementation researcher at Drexel University in Philadelphia, who focuses on issues related to mental health and health equity. "This is kind of one of these counterintuitive, interesting things," Dr. Purtle said. "When you've experienced these shocks and stressors throughout your life, you learn to get really good at coping, bouncing back, and dealing with stuff. For people who haven't had those stressors, when disaster strikes, it can be worse for their mental health because they haven't developed the coping skills that serve them in these contexts. But the literature on this is kind of mixed."

Dr. Purtle added, "I think we also see these mental health impacts following natural disasters more now because the exposures are greater."

Those living on small incomes, as well as minorities, disproportionately experience the negative impacts and environmentally induced mental distress. Researchers say this is due to more fragile overall health, reduced mobility, reduced access to health care, and the inability to purchase goods and services to mitigate the effects of disasters.

Weather Preparedness

Mental health experts say there are no "right" or "wrong" feelings when dealing with the stress that follows natural disasters. One way to feel more confident and in control is to be proactive in preparing for emergencies. Creating a family disaster plan can help.

This starts with becoming aware of the biggest natural disaster risks in your community and staying alert to weather advisories for your area. In case there's ever the need to evacuate, gather essential items in advance and assign responsibilities to each family member so that everyone understands what actions to take in the event of an emergency.

Families should designate one out-of-area emergency contact person. This information should be shared and saved on all cell phones. Texting would likely be unavailable when cell service is down, but there are some messaging apps to consider downloading in advance of a storm, like the walkie-talkie-style Zello, which works over Wi-Fi or your phone's cellular network. Family and friends can join a group on the app, and through the "microphone" button, you can speak to everyone in real time.[59] The *Houston Chronicle* reported that Zello messaging was used during the round-the-clock water-rescue operations of the "Cajun Navy," a volunteer group that assisted in locating and helping those trapped in floodwaters during Hurricane Harvey in Texas in 2017.[60]

FireChat is a free messaging app for both Android and iPhone users that doesn't require Wi-Fi or cellular data. As Bustle.com points out, though, there's a catch: "While this sounds like a dream come true, the one caveat is that you need to create a group to chat with people no more than 210 feet away."[61] FireChat can be particularly useful for close, crowded areas where the cellular signal is weak or nonexistent, like disaster shelters, according to the *Charlotte Observer*. The app uses peer-to-peer Wi-Fi and Bluetooth to keep everyone's phones connected.[62]

Planning for your pets' safety in the event of an emergency is not only a good idea for the animals. It can also alleviate human anxiety levels. The Paw-Boost app allows you to view, share, and report lost and found pets. You can also choose to be notified when a local pet is lost or found near you so you can contact the owner or finder with helpful information.[63]

Practicing your family disaster plan twice a year is advised. This includes going through the routine of evacuating, planning escape routes, and ensuring that information and supplies are readily available to everyone.

Coping Kits for Kids

Taking these precautions can also help children cope with storm anxiety. According to experts at Texas Christian University, weather preparedness can contribute to their sense of safety. Since natural disasters can occur with little notice, it's advised that parents also assemble their children's coping or self-soothing kits to provide outlets for children to calm themselves if triggered by inclement weather.

These kits may include drawing or coloring materials, headphones, music, or favorite books. The Karen Purvis Institute of Child Development at Texas Christian University suggests that parents talk with children about what's in their self-soothing kit, where it's located in the house, and how the child can use it when he or she is feeling anxious.[64]

Preparing and restocking emergency kits in advance is also important. These kits should include, at minimum:

- Water: one gallon per person, per day (three-day supply for evacuation, two-week supply for home)
- Food: nonperishable, easy-to-prepare items (three-day supply for evacuation, two-week supply for home)
- Pet food and essentials (if needed)
- Flashlight
- Battery-powered or hand-crank radio (a NOAA weather radio, if possible)
- Extra batteries
- First-aid kit
- Medications (seven-day supply) and medical items
- Multipurpose tool
- Sanitation and personal hygiene items
- Copies of personal documents (medication list and pertinent medical information, proof of address, deed/lease to home, passports, birth certificates, insurance policies)
- Cell phones with chargers
- Family and emergency contact information
- Extra cash
- Emergency blanket

(Source: FEMA, Ready.Gov)

Restoring Earth to Heal

A garden is a place of healing to the soul.
—Harriet Beecher Stowe, 1855[65]

Once brimming with vegetables and flowers, the community garden at Beach Forty-First Street Houses in Far Rockaway, New York, was washed away in

2012 by Hurricane Sandy. After the floodwaters eventually receded, only weeds sprouted in the thirty garden plots of the New York City public housing property. As the Queens neighborhood repaired damages to their homes and apartments, nearby Housing Authority residents longed to have their beloved garden restored as well. They missed planting flowers and vegetables, hoeing the rich soil, and smelling the garden's grass and herbs.

"People have an innate need to connect with nature, which intensifies following crippling disasters," said Keith Tidball, director of the New York Extension Disaster Education Network at Cornell University. Tidball researched global disaster zones to understand better how people heal themselves through restoring nature.[66]

Tidball—along with other researchers from Cornell University, social scientists from the US Forest Service, and the TKF Foundation, a nonprofit that advocates for and supports the creation of healing green spaces—joined forces with community volunteers on a mission to revive the storm-decimated plant plots into a new "Hurricane Healing Garden."

In a 2015 documentary short about the project, residents shared their heart-wrenching firsthand accounts of their terrifying experiences living through the storm and its aftermath. Despite their different ages and backgrounds, they shared the common sentiment that the storm-destroyed green space had been a place of joy and tranquility.[67]

"When you go through the gate and you come over here," one resident said in the film, entering the patch of land where the garden was situated, "there's a feeling that comes over you, and you're going to be at peace."[68]

Tidball believes that after natural disasters, humans have an intrinsic desire to reconnect with nature. It's part of the coping process. "Remembering that affinity and the urge to create restorative environments . . . demonstrates resilience,"[69] he said, describing this drive as *urgent biophilia*. (*Biophilia* is the idea that humans possess an innate tendency to interact or be closely associated with other forms of life in nature.)[70]

If so, projects like the Beach Forty-First Street garden will become even more essential in the wake of increased natural disasters due to climate change. "It's part of a community's long-term recovery," Tidball said in an interview with NatureSacred.org.[71]

Trusting Nature Again

Some experts suggest that after experiencing a devastating weather event, survivors find it difficult to trust nature again. In the 2019 book *Aftershocks of Disaster: Puerto Rico Before and After the Storm*, coeditor Yarimar Bonilla, a political anthropologist specializing in sovereignty, citizenship, and race across the Americas, shares a conversation with journalist and author Naomi Klein.[72] Klein has written extensively about Puerto Rico and climate change herself, notably for the *Intercept* and in her book *The Battle for Paradise*.

In the aftermath of Hurricane Maria, Klein tells Bonilla, "One of the most moving experiences I had was at this farm school in Orocovis, Puerto Rico."[73] Klein described the agro-ecology school, where she observed students in the fields post-Maria as they planted and harvested crops in the "juxtaposing setting of hurricane devastation all around them."

Klein recalled the conversations she had with the school's director, who told her that "this work gave its students agency after Maria." In watching the students in the gardens, she observed that despite the disaster, the children were "happy and energized by their work, which was helping to feed their families."[74]

Klein explained that living through a monster hurricane like Maria can make a person feel like "the natural world has turned against you. Learning to trust the natural world again, as a source of strength and substance, reminds us that we are part of a web of life, that the land can support life."

Earthly Mindfulness

Gardening, planting, and engaging with the outdoor environment can also promote healing from stress after disasters through an additional practice that's beneficial to emotional health: *mindfulness*, or being fully present in the moment.

Research from the University of Massachusetts indicates mindfulness may offer relief from post-traumatic symptoms such as anxiety, sleep disturbance, and difficulty concentrating.[75] Their study evaluated the effectiveness of an eight-week program implementing meditation, body awareness, yoga, and exploration of patterns of behavior, thinking, feeling, and action. Researchers concluded, "we found an improvement in PTSD symptoms and an increase in mindfulness that could be associated with an improvement in spiritual well-being."[76] Many other mindfulness-based treatments for post-traumatic stress disorder (PTSD) have also emerged as promising adjunctive or alternative intervention approaches.[77]

Dr. Amy Reyer teaches mindful meditation out of her Hudson County, New York, studio and remotely to clients nationally. She has experience addressing patients' anxiety following unforeseen adverse events.

"Worrying over the potential loss of loved ones or being separated from them—it turns out that that kind of fear can be bad for our health. Especially if we don't have resources for how to deal with it," Dr. Reyer said.

Dr. Reyer gave the example of one type of mediation, called *loving-kindness*, which involves repeating phrases of positive intention for self and others. "It's a way of being with people you love and kind of calling them to mind and heart," Reyer explained.

In a follow-up with veterans suffering from PTSD who recently participated in a Seattle-based twelve-week course learning and practicing the loving-kindness meditation, researchers found that "Overall, this meditation appeared safe and acceptable and was associated with reduced symptoms of PTSD and depression."[78]

How to Meditate

Dr. Jack Kornfield trained as a Buddhist monk in the monasteries of Thailand, India, and Burma. He has taught meditation internationally since 1974 and is one of the key teachers to introduce Buddhist mindfulness practice to the West.[79] On his website, jackkornfield.com, he offers a sample of a loving-kindness meditation along with instructions.

Dr. Amy Reyer, who has been researching, practicing, coaching, and writing about mindfulness-based personal and professional transformation for over ten years (artoflivingslowly.com), explains the science behind successful meditation. "There's a feeling that we get through the practice that activates the brain in the way we feel a sense of warmth and love and connection to another person or a pet; the practice activates in our mind and body that same way," she says.

Meditation on Loving Kindness

Breathe gently and recite inwardly the following traditional phrases directed toward your own well-being. You begin with yourself because without loving yourself; it is difficult to love others.

> May I be filled with loving-kindness.
> May I be safe from inner and outer dangers.
> May I be well in body and mind.
> May I be at ease and happy.

As you repeat these phrases, picture yourself as you are now, and hold that image in the heart of loving-kindness.

Or perhaps you will find it easier to picture yourself as a young and beloved child. Adjust the words and images in any way you wish.

Create the exact phrases that best open your heart of kindness. Repeat these phrases over and over again, letting the feelings permeate your body and mind.

Practice this meditation for a number of weeks until the sense of loving-kindness for yourself grows.[80]

(Source: jackkornfield.com)

Gratitude

Practicing gratitude may also help reduce stress and anxiety following disasters. Researchers studied the role gratitude had in improving the emotional well-being of earthquake survivors in the Ya'an region of southwestern China following the natural disaster that occurred there in 2013.[81] They uncovered that social support along with gratitude had a stable and positive effect on the survivors a year after the disaster. Research also pointed to a "decreased likelihood of PTSD"[82] even three and a half years after the earthquake.[83]

Gratitude helps people feel more positive emotions, relish good experiences, improve their health, deal with adversity, and build strong relationships.[84] The UCLA Mindful Awareness Research Center stated that gratitude actually "changes the neural structures in the brain, and makes us feel happier and more content."[85]

Giving thanks and acknowledging appreciation can also be done through keeping a journal—by writing down things daily we feel grateful for. The Cleveland Clinic offers these tips to get started:[86]

1. **Write when it works best for you.** Some evidence suggests that writing in your gratitude journal right before bed may help you sleep better by forcing you to stop worrying and shift focus to good things. However, you may find writing in the morning or jotting down thoughts as they happen works best. Find whenever you're most likely to do it.

2. **Keep it simple when beginning.** Start off by listing three things in each journal entry. If you're struggling, first ask yourself if you're grateful for food, hot water, a roof over your head. Detail what you already have, and you may quickly discover things that you often take for granted.

3. **Look for the good throughout your day.** Instead of dwelling on the negative, you can consciously shift your focus to positive thoughts with the help of a gratitude journal. This enables you to be more present, mindful, and aware of the goodness in your life as you're experiencing it.

4. **The more details, the better!** Once you get the gist of it, look for the little details—this helps re-create memories and may even inspire a mind-body response. Ever laugh out loud while recalling a funny story? As you recall a special moment, you ignite the pleasure of reliving it.

The practice of gratitude can also be expressed through meditation. Dr. Reyer says a rewarding way to incorporate the loving-kindness meditation is by using it to focus on others rather than yourself. In this type of meditation, you bestow benevolent and loving energy, feelings of goodwill, kindness, and warmth.[87]

If you're new to mediation, Dr. Reyer advises beginning gradually. "I would say a maximum of fifteen minutes a day for starters," Dr. Reyer said.

In this meditation, "You're sending gratitude and well wishes to loved ones, friends, and it just feels very good." Dr. Reyer suggests breaking up the time into five-minute increments, three times a day. "That's a practice that over two weeks, you'd actually start to notice some changes in your overall well-being."

3.

CITIES OF HEAT

During the summer of 2019, a deadly heat wave hit Europe not once but twice, with devastating consequences. In Paris, temperatures soared to an all-time record of 109 degrees. Fifteen hundred people died. According to the French Health Minister, half of them were seventy-five years old or older.[1] This grim statistic underscores the severe risk heat poses on vulnerable residents living in urban areas.

It's not just in Paris. As urban populations expand globally, tall buildings, black asphalt, and less green space make the world's cities hotbeds for heat. Extreme heat causes more deaths in US cities than all other weather events combined.[2]

Scorching hot days are an overt hazard. Equally as concerning are oppressively steamy summer nights, which have increased in frequency due to climate change. When temperatures don't fall below 80 degrees Fahrenheit at night, the human body isn't able to recover from the impacts of extreme heat.[3] These negative health impacts are amplified by the urban heat island effect, which keeps cities substantially hotter than rural areas.

The elderly—particularly those who live alone in high-rise, heat-trapping older apartments—young children, and residents of low-income communities are especially susceptible when city temperatures soar.

There are personal health precautions to be aware of and implement before excessive heat arrives. On a communal scale, planting more trees adds natural cooling, improves air quality, and boosts mental health. Lush landscapes brimming with rooftop greenery can help structures stay cooler inside and offer additional outdoor space. There are also ways to reduce latent heat trapped inside city dwellings.

Sizzling Cityscapes

Individual city skylines are distinct. But when viewed from above, cityscapes reflect certain similarities. If you were to paint a picture from an aerial perspective, you would likely find yourself choosing colors like dark brown, charcoal gray, and faded black to reflect the dull, murky hues of skyscrapers, streets, and rooftops.

These lackluster shades also play a role in turning up the thermostat in cities on hot days. That's because the sun's ultraviolet light interacts with these colors differently than it does with lighter shades. Cement and steel buildings and brick and asphalt streets soak up the heat transmitted in UV waves rather than reflect it. That keeps all surfaces hotter. In meteorology, this is known as the *albedo effect*, the measure of how much light that hits a surface is reflected without being absorbed.[4] The albedo effect is why wearing light-colored clothing in the summer will keep you cooler, and another reason why city streets seem to bake in the hot sun.

Some cities have been getting warmer each year at a faster rate. According to Climate Central (an independent organization of scientists and journalists researching and reporting the facts about our changing climate and its impact on the public), from 1970 to 2018, temperatures for southwestern American cities like Las Vegas, El Paso, Tucson, and Phoenix have seen the thermometer's most significant upticks, with average increases of 4.3°F in the nearly four decades studied.[5]

Urban Heat Islands

Why are cities especially dangerous when it comes to extreme heat? The answer is the urban heat island effect. Even with equal hours of sun, the temperature reported in a city compared to a nearby rural area can be ten to fifteen degrees warmer on the same day at the same time!

There are several reasons for this disparity. Rural areas benefit from nature's greenery that acts as a cooling system through the process of *transpiration*. Deep in the Earth's soil, plants and trees absorb water, pulling water up through their stem or trunk and then pushing that moisture outward to quench leaves. Through tiny plant pores, water vapor is released back into the air. This natural moisture flow provides a cooling influence on the surrounding air—just as when sweat evaporates off your skin, your body feels cooler.

Contrast this with artificial surfaces covered with solid, dense, dark building materials that can't suck up water of any kind. For different reasons, they were designed to be that way. Rather than architectural distinction, streets were designed to roll rainwater off into drains. Curbs with slightly depressed gutter sections and streets with peaked centers (called street crowns) were raised from each side by about 1 to 2 percent. This upward curve allows stormwater to flow into drains. Otherwise, when it rains, the water would stop traffic flow and make driving dangerous. The intended objective was to prevent hydroplaning, visibility impairment from splashback, and car engine stalls.[6]

In a densely populated city like New York, these impervious surfaces cover approximately 72 percent of its 305 square miles in land area.[7] That means minimal natural cooling.

A yearlong study of Dallas, Texas, released in 2017, showed that a full 35 percent of the city was covered by grass-free, water-proof surfaces, like parking lots, roads, and buildings. The hottest spots within the city measured an average high of 101°F and a low of nearly 80°F for five whole months of the year.[8]

Warm Summer Nights

When the sun finally sets after a long, hot summer day, the city itself seems to exhale a deep sigh of stifling air. But lately, nightfall has been providing little relief. That's according to the findings of a study released in October 2020, in which University of Exeter scientists studied temperatures from 1983 to 2017. For those thirty-four years, they found that nighttime warming was more than twice as apparent as daytime warming.[9]

Health researchers say that exposure to nighttime heat, particularly when preceded by a hot day, contributes to a multitude of health consequences, including, at worst, heat-related death. This dire consequence occurred most often in people who were also at greatest risk for stroke or heart diseases.[10]

Summer heat can additionally cause breathing difficulties for those living in cities where the air quality is already poor due to increased ozone levels in the atmosphere.

Good and Bad Ozone

Ozone is a gas composed of three atoms of oxygen (O_3) found in the Earth's upper atmosphere and ground levels. Ozone can be good for the environment or bad for your health, depending on where it's located.[11] Stratospheric ozone, the "good" kind, naturally occurs in the upper atmosphere, where it forms a protective layer that shields us from the sun's harmful ultraviolet rays. Ground-level ozone found at the Earth's surface is "bad" because it can trigger various health problems.

Ground-level ozone forms through chemical reactions among nitrogen dioxide, carbon-based substances like methane or other compounds, and the sun. These reactions can be triggered by emissions from cars, power or industrial plants, or refineries.[12]

The bright, hot sun mixes with pollutants to create a thick, unhealthy haze, often on the muggiest days of summer—making the air tough to breathe. Health effects include irritated throats, coughing, inflamed airways, and chest pain. High ozone levels can reduce lung function and damage lung tissue, worsen bronchitis, emphysema, and asthma, and in severe cases, send you to the ER.

The Most Vulnerable

When you think of children playing outside their homes or apartments, you may envision kids riding bikes along leafy suburban streets. But in the US, roughly 81.2 percent of children live in urban areas.[13] In fact, children make up a substantial share of the urban population of American metropolises—most notably Salt Lake City (29.4 percent), followed by four Texas metro areas—Houston, Dallas, Fort Worth, and San Antonio. Memphis, Atlanta, Phoenix, Raleigh, and Indianapolis round out the top ten urban areas with the most child-age residents. Note that eight out of these ten large metros are located in the hot Sun Belt states.[14]

Children are more vulnerable to heat than adults—in ways you may not realize. Their smaller bodies have thinner skin and more of it per pound of body weight. They hold less fluid internally, which means that water loss—as in dehydration—can significantly affect them. Their body temperature rises three to five times faster than adults—making the impact of heat on their tiny bodies almost immediate.[15]

Early signs of dehydration in children include fatigue, thirst, dry lips and tongue, lack of energy, and feeling overheated. Children often don't think to drink water until they're thirsty, but that may be too late. Medical experts say thirst may not kick in for a child until they've already lost 2 percent of their body weight as sweat.[16]

Safety precautions parents can take for kids during heat waves, according to the CDC, are to limit strenuous outdoor activity for young children and make sure they take regular water breaks when playing outside. They advise that, if possible, parents should bring outdoor activities inside on extremely hot days.

"Kids are at greater risk, it turns out, because they're not just little adults," Dr. Mona Sarfaty points out. "Their surface area to mass is different than adults, and they spend a whole lot more time outside."

Dr. Sarfaty is not just an expert in family medicine; she is also the Director of the Program on Climate and Health in the Center for Climate Change Communication at George Mason University. Dr. Sarfaty says summers today are warmer than the ones that parents may remember themselves experiencing as kids.

"The parents who grew up in a different time probably don't realize how much of a risk heat is for their children," said Dr. Sarfaty.

The *New York Times* published an interactive visualization tool that illustrates Dr. Sarfaty's point. In their August 2018 article "How Much Hotter Is Your Hometown Than When You Were Born?"[17] there's an interactive prompt to enter your birthplace and birth year. Then you're given past and present weather data to gauge the difference between then and now. For example, if you were born in New York City in 1969, findings illustrate that you can expect to see five more days when high temperatures reach 90 degrees than in the year when you were born.[18]

Aging in Heat

Another group vulnerable to excessive heat are the elderly, those aged sixty-five and above. As we age, our body's natural ability to cool slows, which is imperative during a heat wave. Older people also may be on medications that make them more sensitive to extreme heat. Diuretics, sedatives, tranquilizers, and certain heart and blood pressure drugs can hinder sweating.

Lifestyle factors can contribute to risk, including extremely hot living quarters only accessible by stairs, lack of transportation, overdressing, visiting crowded places, and not understanding how to respond to weather conditions.[19] Heat concerns for seniors are growing because the population of older Americans is increasing. It's estimated that by 2040, one in five Americans will be over the age of sixty-five.[20] Cities with strong senior populations are not just in Florida and Arizona. In Pittsburgh, 18.3 percent of the population is over sixty-five—which is 26 percent higher than the national average.[21]

Living Alone

Older adults living in cities are more likely to live by themselves.[22] Globally, this is more common in countries with advanced economies, according to Pew Research. People tend to have fewer children and have them later in life; they are also more likely to live well beyond their childbearing years. Governments in wealthier countries also may offer financial assistance or health care benefits for retired adults. This makes it more affordable for older people to stay in their own homes.[23]

This is particularly true for Americans. Whereas seniors in other countries are more likely to reside with extended family, in the US, 27 percent of adults ages sixty and older live alone, compared with 16 percent of adults in other countries studied around the globe.[24]

Solo Living

You may have heard the expression that being "alone" and feeling "lonely" are not the same thing. But for seniors who live in cities, particularly if they aren't homeowners, the isolation and transiency may lead to feelings of disconnection from their community.[25] According to research presented in the article "What's the World's Loneliest City?" in the *Guardian*, older urbanites living on their own may have a more challenging time making and keeping permanent social connections.[26] The article cites various reasons: "Cities are filled with lots of renters—such as London, which is expected to have 60 percent of residents renting by 2025—have greater transience and potentially lower community engagement."[27]

Loneliness and seclusion for seniors in urban areas can cause them to be at greater risk for physical health problems that can be exacerbated by extreme heat. In his research for the National Institute on Aging, Dr. Steve Cole, Director of the Social Genomics Core Laboratory at the University of California, Los Angeles, described loneliness as "a fertilizer for other diseases."[28]

"The biology of loneliness can accelerate the buildup of plaque in arteries, help cancer cells grow and spread, and promote inflammation in the brain leading to Alzheimer's disease. Loneliness promotes several different types of wear and tear on the body," Cole said.[29]

The Two Solitudes

The most severe consequences can occur for the elderly living alone in city apartments with no air-conditioning. A tragic example is found during the record heat wave in Canada in July of 2018 when temperatures soared past 95 degrees for days.[30] The heat wave dealt a devastating blow to Quebec: 74 deaths were linked to excessive heat.[31]

Marco Chown Oved, an investigative reporter for the *Toronto Star*, analyzed the 2018 heat wave in the article "Life and Death Under the Dome." He pointed out that while sixty-six people in Montreal died of heat-related illness, approximately 200 kilometers away in Ottawa, the human death toll played out differently.[32]

"The heat and humidity in Ottawa were just as brutal and lasted just as long. But according to the provincial coroner, no one died. How could it be that so many died in Quebec while there was no heat-related death in Ontario?" Chown Oved wrote in his article, "It's a question that risks exacerbating the divide between Canada's two solitudes, but also one that will become vital for people from coast to coast to coast in the years to come."[33]

The "two solitudes" Chown Oved refers to in his report are the cultural and language distinctions between Canada's dual largest provinces, divided by the Ottawa River. But as Chown Oved goes on to explain, there are other disparities that factor in when it comes to differentiating between Quebec's and Ontario's survival rates in a dangerous heat wave. Quebec residents are less likely to have air-conditioning than those living in Ontario. In fact, only 53 percent of Quebec households have air-conditioning, while 83 percent of Ontarians have some form of AC at home.[34]

This may have been a substantive factor in the varying death toll between the provinces. Later, a report from Quebec's public health authority provided further details. Most of the deaths in Quebec were men who lived alone;[35] many had underlying physical or mental health conditions. Almost all who died were over sixty years old. And the majority of deaths "happened in densely built-up parts of the city—neighborhoods where heat-trapping concrete and sparse vegetation elevated the temperatures by around 5°C to 10°C."[36] These were the hottest and despairingly deadliest spots of Montreal and Quebec City's urban heat islands.

Mapping Out City Heat

In order to better understand the urban heat island effect and how to mitigate its health impacts, US citizen scientists are helping NOAA's Climate Program Office (CPO), the National Integrated Heat Health Information System (NIHHIS), and the Centers for Disease Control (CDC) map city heat and collect data annually.

Here's how the program works: during one of the hottest days of the year in each of the designated US cities, volunteers drive predetermined routes in the morning, afternoon, and evening with custom-engineered heat sensors mounted on their own cars. Once per second, the sensors record temperature and humidity and the exact time and location along the route.

Using the data collected, the team produces detailed urban heat maps that help officials and community groups identify specific neighborhoods vulnerable to extreme heat.[37] This may seem like a tedious undertaking, but every measurement matters when it comes to reducing the number of people dying from heat-related illnesses. Yale University scientists found that heat wave mortality risk increased 2.5 percent for every 1°F increase in heat wave intensity and 0.38 percent for every one-day increase in heat wave duration.[38]

Tree Tracking

This data is even more valuable when combined with a detailed analysis of where trees stand tall and provide shade known as *urban tree canopy* (UTC). These locations can be changeable within each city. While trees may live for hundreds of years, their roots may not remain planted deep in the ground at precisely their original origin. Or trees might have been removed after storms or in advance of land development.

In fact, city trees are rapidly losing ground. Tree cover in US urban areas is declining at about 175,000 acres per year, which corresponds to approximately 36 million trees per year.[39] According to americanforest.org, the US is losing one tree for every two planted trees due to extreme weather events, insects, and diseases; trees being removed for buildings; and improper planting practices. Nationally, if we stay on this path, we're facing a projected loss in urban tree cover of 8.3 percent by 2060.[40]

For example, in Louisville, Kentucky, recent studies estimate the city is losing an average of fifty-four thousand trees a year.[41] To slow down this tree loss rate, new legislation was passed in April 2020 that will require development sites in Louisville with 50 percent to 100 percent existing tree canopy coverage to maintain at least 20 percent of that canopy.[42]

Trees are crucial to reducing city heat, as they act as a natural air-conditioner. According to Canopy.org, "The evaporation from a single tree can produce the cooling effect of ten room-size, residential air conditioners operating twenty hours a day."[43]

The Inequity of Heat

In dozens of major US cities, low-income neighborhoods endure the hardship of heat disproportionately to those living in wealthier sections.[44] According to a joint investigation by NPR and the University of Maryland's Howard Center for Investigative Journalism, "Those exposed to that extra heat are often a city's most vulnerable: the poorest and, our data show, disproportionately people of color."[45]

In Miami, Florida, city demographics reflect that the population is about 65 percent Hispanic, followed by approximately 18 percent Black.[46] Haitians and African Americans are more likely than any other major ethnic group in

Miami to live in poverty, with poverty rates of 45 percent and 44 percent, respectively.[47]

While most Floridians do have air-conditioning, those who don't often live in low-income areas. This is problematic even for Floridians who are accustomed to year-round tropical climates. That's because hot weather in south Florida is more intense and lasts longer than it used to. For example, in 1970, Miami saw an average of eight consecutive days of 90-plus degree temperatures. Research shows that in 2019 that number jumped up to around twenty-five straight days of 90-plus heat.[48]

Little Haiti and Heat

Before there was the city of Miami, though, there was the community of Lemon City, along the shores of Biscayne Bay in south Florida. Named for its lush lemon trees, the community's roots go back to the 1870s. Today, though, the area is better known as Little Haiti.

The name change occurred in the 1980s, as waves of Haitian immigrants fleeing dictatorship sought refuge in Miami and settled in this area, as it was the most affordable. The neighborhood is culturally vibrant, abundant with murals, music, food, and art. Little Haiti is home to approximately 30,000 residents, about 75 percent of whom are Black; the community is considered one of the city's historically low-income areas, with approximately 47 percent of its residents living below the federal poverty line.[49]

Most of the neighborhood's residents are renters. Only about a quarter of Little Haiti's housing units are owner-occupied, and many homes were built after 2000. About 53 percent of the housing was built before 1960.[50] These older homes are hotbeds for heat.

"Oh my gosh. I'll never forget it. We walked in, and it was incredibly hot for the family of five living there," recalls Dr. Cheryl Holder about a home visit to a patient in Little Haiti. Dr. Holder has dedicated her medical career to treating underserved populations. She is the founder and co-chair of Florida Clinicians for Climate Action.

"These little apartments have one window. . . . The mom had one little fan that she had over at the baby, who was fast asleep. She was giving us the fan, and we were like, no, that's okay." The older home didn't have air-conditioning.

Landlords aren't required to provide air-conditioning, according to landlord–tenant laws, as it's considered an amenity.[51] For those living under federal assistance, the Department of Housing and Urban Development (HUD) does not mandate air-conditioning in public housing. "Federal housing has heating in Miami, but no air-conditioning," Dr. Holder said, explaining the paradox.

Dr. Holder also said she'd visited some patients in their home who may have a single air conditioner but don't want to run it due to the electricity cost. "Central air isn't everywhere," Dr. Holder said. "They may have one room with an AC, but they can't afford to use it."

According to an April 2020 report from the University of California, Berkeley, a ten-year energy spending analysis in the US found that while families with annual incomes of $50,000[52] or greater spend on average 3 percent of their income on electricity, those earning less than $10,000 a year allocate more than ten times that amount toward their cost of cooling.[53] That's a significant disparity. It means choosing to use air-conditioning may cost you the funds for other necessities.

"One of my patients' households I went to visit was a woman living alone, sheltering in place for coronavirus [in 2020]," said Dr. Holder. "Her one window AC unit broke. She has diabetes....The *Miami Herald* interviewed me at the time, and I told them about her. Thankfully, a reader of that article ended up donating an air conditioner to my patient."

Federal AC Help

The Home Energy Assistance Program (HEAP) is a federally funded program that helps low-income people pay for the cost of heating or cooling their homes. Those eligible may receive one Cooling Assistance benefit for the installation or purchase of a fan or an air conditioner to help cool your home. Each applicant household can only accept one benefit for Cooling Assistance. To be directed to your local HEAP provider, call (800) 282-0880.[54]

Trees of Prosperity

Instead of looking left or right while standing in the heart of a neighborhood, gaze upward, as that may prove to be a more valuable vantage point. Do you find yourself situated beneath a canopy of lush, leafy trees?

According to research, the wealthier a community is, the more trees they have planted.[55] Bountiful branches mean more shade from the sun and greater natural cooling. These science-based findings originally came out in 2008 but drew more attention when Tim De Chant, the author of the blog *Per Square Mile*, broke down the interesting numbers related to trees and money in a popular post in 2012: "For every 1 percent increase in per capita income, the demand for forest cover increased by 1.76 percent. But when income dropped by the same amount, demand *decreased* by 1.26 percent. That's a pretty tight correlation. . . . Wealthier cities can afford more trees, both on private and public property. The well-to-do can afford larger lots, which in turn can support more trees."[56]

Leafy canopies conserve moisture, slow the wind, reduce traffic noise, and as discussed earlier in this chapter, provide cooling shade from the hot summer sun. According to canopy.org, "Trees are natural air conditioners." Shading and evaporational cooling from trees can cut home AC use almost in half.[57]

The answer to the age-old question "Does money grow on trees?" may very well be that the trees are actually as valuable as money itself.

Take Palo Alto, California, for example. The city's name is credited to a tall, millennium-old redwood that now looms over the Caltrain tracks.[58] One of the wealthiest cities in America, home to tech company giants, Palo Alto is also rife with another kind of green: plentiful tree canopy coverage, which ensures that summer temperatures are at least six to eight degrees lower than in comparable neighborhoods without trees.

In a 2015 study published in *PLoS One*, Boise State University economist Michail Fragkias and others explored the relationship between the number of city trees and overall income levels in seven US cities: Baltimore; Los Angeles; New York City; Philadelphia; Raleigh; Sacramento, California; and Washington, DC.[59] The study's authors suggest that the relationship between urban tree cover and income may result from a *feedback loop* where more significant tree cover increases property values, which in turn attracts households with higher incomes.[60] Similarly, areas with fewer trees tend to have less costly homes. Those residents may have less access to resources because they are renters or on fixed incomes.[61]

Cities are typically where the most jobs are and where public transit systems are more accessible. Moving out of the city and into a greener, more rural community can potentially lower housing costs, but it can then increase commuting costs.[62] That, along with other factors, makes relocation challenging. "Those living in low-income communities have fewer choices in where to live, in this country. There's not a lot of affordable housing," said Laurie Schoeman, the senior director at Enterprise Community Partners, a national nonprofit working to preserve and protect affordable housing from natural hazards and a changing climate.

Schoeman's expertise is in high demand as the correlations between environmental risk and low-income housing availability grow. In metropolitan areas like Boston, Atlantic City, and New York City, research from Climate Central shows that each city could have thousands of low-income housing units exposed to chronic coastal flooding by 2050.[63] According to *Bloomberg Reports*, these environmental threats only exacerbate America's existing problem of having just thirty-five units available for every one hundred extremely low-income renters.[64]

"There are few options for folks living on a low income. So, it's not like they're going to be able to leave a community where they have Section Eight housing or subsidized housing and go to another community and find the same housing. It may not exist," Schoeman said.

Cooling Inequity

Digging deeper into the disparity of urban tree coverage within cities, another study published in 2020 by researchers from Portland State University, the Science Museum of Virginia, and Virginia Commonwealth University offered additional insight.[65]

The authors identified the hottest spots within the urban heat islands of each of the 108 cities they analyzed. In these locations during summer heat waves, surface-temperature differences in high vs. low UTC areas varied by as much as seventeen degrees.

The researchers found that 94 percent of those superhot areas matched up geographically with zones that were initially designated (in the 1930s) by the US government as "hazardous" in terms of presenting the greatest risk for banks looking to finance home mortgages. These neighborhoods, which received the lowest grade for loans by the government at that time, in comparison to other wealthier sections of each city, were then populated by predominantly poor African Americans, immigrants, and other minorities.[66]

This practice was later referred to as *redlining*. Dr. Gwen Sharp, professor of sociology at Nevada State College, writes in the Society Pages that "red-lining made it difficult for Blacks (and some white ethnics) to buy homes."[67] She says racial discrimination led to "Blacks being unable to buy homes outside of these neighborhoods. As a result, African Americans were disproportionately barred from one of the major avenues to acquiring wealth-building equity through homeownership."

While the practice of redlining was legally banned in 1968,[68] researchers have determined that even today, these same locations within cities—those neighborhoods covered in the highest amounts of impermeable concrete and asphalt surfaces and having the lowest amount of tree canopy—are inhabited by primarily low-income, minority residents. These are precisely the sections of urban areas that are the most vulnerable to extreme heat.

Additional research indicates that communities of color are three times more likely than white communities to live in "nature deprived places."[69]

Thickening up leaf canopy by adding more trees to these parts of cities may be one way to help reduce heat-health risks. According to americanforests.org, approximately 1,200 heat-related deaths and countless heat-related illnesses are prevented because of trees each year.[70] And trees offer plenty of other health benefits as well, removing pollutants by absorbing them through the pores on the leaf surface. Particulates are trapped and filtered by leaves, stems, and twigs and washed to the ground by rainfall.[71] This helps reduce the dust, ash, pollen, and smoke that swirls in city air, damaging human lungs.

Trees also promote mental health and wellness for urban residents. Researchers in London found that a higher street tree density of one tree per kilometer was associated with 1.18 fewer antidepressant prescriptions per thousand people.[72]

In order to get more trees planted where they're needed, American Forests, the oldest national nonprofit conservation organization in the United States, created a tool to help community activists, urban foresters, and city planners. It's called a *Tree Equity Score*. This is calculated by evaluating neighborhood data, including existing tree canopies and street-level temperature readings along with population density, income, age, race, and employment status.

They combine the metrics into a single score between 0 and 100. A score of 100 means that a neighborhood has achieved Tree Equity.[73] This means they have enough canopy cover for residents to reap the health, economic, and other benefits that trees provide.[74] American Forests refers to this effort as a "moral imperative, not just an environmental justice issue."[75]

Planting more trees around the world is an international effort. In August 2020, the US launched the first American chapter of One Trillion Trees, a global initiative of corporations, nonprofit organizations, and governments pledging to conserve, restore, and grow more than 855 million trees by 2030.[76]

Keep It Cool on Top

Another way to mitigate urban heat risk—instead of going bottom-up, start top-down. The concept is called *cool roofs*. Through the use of lighter-colored surfaces or special coatings to reflect more of the sun's heat, these roofs improve building efficiency by reducing cooling costs and offsetting carbon emissions.[77]

When the sun shines on a dark rooftop on a hot day, most of its UV rays strike a rooftop and then reflect into the sky. But not all of it. Some of that solar energy or radiation is absorbed as heat in the roof. This not only heats the top tier but everything below it and the surrounding air.

Traditional dark roofs strongly absorb this sunlight. This increases energy use in air-conditioned buildings and makes non-air-conditioned buildings less comfortable. Hot dark roofs aggravate urban heat islands by warming the air flowing over the top and contributing to global warming by radiating heat into the atmosphere.[78]

On a typical summer afternoon, a cool-colored roof that reflects 35 percent of sunlight will stay about 22°F cooler than a traditional roof that looks the same but reflects only 10 percent of the sun.[79] The Environmental Protection Agency (EPA) states that the temperature of roof surfaces can be 50-90°F higher than the air temperature!

In 2015, L.A. became the first major city to mandate cool roofs for new residential construction projects. Dallas and other major cities quickly followed suit. Flat or low-sloped roofs were first transformed by coating them white. However, there are now "cool color" products on the market, darker-colored pigments that are still highly reflective in the near-infrared (non-visible) portion of the solar spectrum.

The combination of tree planting and preservation with cool roofing and paving materials across Dallas was found to yield greater health benefits for the city than any individual heat mitigation effort. These combined strategies were found to reduce the number of deaths from hot weather by more than 20 percent.[80]

The cooling concept is also being used on the streets themselves. The City of Phoenix Street Transportation Department selected portions of eight neighborhoods and one city park to receive cool pavement treatment as part of a pilot project.[81]

Another way to reduce roof heat is to turn roofs "green." Following the heat wave of 1995, the city of Chicago installed 359 green roofs—covered partially or entirely by plants, trees, and greenery.[82] Look for more green and cool roofs to pop up in the years ahead!

For City Residents:
Tips to Keep Your Apartment/Homes Cooler

Suppose you don't have an AC or have one and would like to use it less. Here are some tips to stay as cool as possible in the summer heat:

- **Open Inside Doors:** keep bedroom and bathroom doors open or ajar to circulate air. Only close doors for rooms you don't use. When you keep doors closed, the airflow into the room is drastically reduced, and the space becomes pressurized, forcing all the cool air out.[83]

- **Open Windows Nightly:** if you feel safe to do so—letting a breeze in at night is a good idea when temperatures are at their low point. Close them during the day to trap this cooler air inside and to keep humidity out.[84]

- **Replace Your Bulbs:** energy-efficient LED bulbs generate less heat and use less electricity than traditional ones.

- **Exhaust Fans:** These kitchen vents above stoves are designed to pull cooking smells and smoke out. They can help suck warm air out as well.

- **Add Plants:** greenery can help cool down a room and improve air quality

- **Power Down Devices:** electronic gear like laptops and small appliances emit heat and use electricity when plugged in, even when off.

- **Close Blinds:** home repair experts say up to 30 percent of unwanted heat comes through sunny windows. Shades or closed curtains can save up to 7 percent on bills and lower indoor temperatures by up to twenty degrees.

- **Use Energy Efficient Appliances:** this can help to lighten the load on the electric grid during heat waves.

4.

VIRUSES AND INFECTIOUS DISEASES— FROM CORONA TO LYME

There's a delicate balance achieved in nature through a healthy, biodiverse ecosystem. When a rich variety of plants and animals interact—each within their own distinct role, yet working together in their shared environment— these species are productive and prosper universally.

Rising global temperatures, increased precipitation, and human intrusion on wildlife habitats may unintentionally collude to tip the scales of nature's perfect paradigm. The result: a greater risk for human illnesses—like future novel coronaviruses, vector and waterborne diseases, and even flesh-eating bacteria.

How can these pathogens be transmitted to humans? One method is in the tainted bite of the ubiquitous mosquito. This pesky creature can carry blood infested with viruses like West Nile or Zika. For humans, this means one infectious bug encounter can result in severe health consequences, ranging from neurological problems to congenital disabilities.

As with many other dangers, minorities and those living in low-income communities are at greater risk. In this instance, living near abandoned buildings and accumulating trash can contribute to residents being more susceptible to infection, as these conditions attract mosquitoes. Commercial pest control can be effective, but it's also expensive.

   Current global warming scenarios suggest greater transmission opportunities for vector and waterborne diseases to be transmitted to humans, but certain preventative measures can help to mitigate the health risks.

More Pandemics Ahead

The novel coronavirus outbreak that began in China in 2019 was declared a worldwide pandemic by the World Health Organization on March 11, 2020.[1] In the months that followed, fear and panic swept the globe. Daily news reports indicated more people were becoming infected, and the death toll was rising.

   In September 2020, in the journal *Cell*, Dr. Anthony Fauci, the nation's leading infectious disease expert, and Dr. David Morens of the National Institutes of Health evaluated the various factors that led to the pandemic in a report titled "Emerging Pandemic Diseases: How We Got to COVID-19." The doctors explained that hazardous exposure to animals through unsanitary conditions, international travel, and growing population density all contributed to the spread of dangerous pathogens: "COVID-19 is among the most vivid wake-up calls in over a century. It should force us to begin to think in earnest and collectively about living in more thoughtful and creative harmony with nature, even as we plan for nature's inevitable and always unexpected, surprises."[2]

   The paper by Drs. Fauci and Morens concluded that the COVID-19 pandemic is "yet another reminder that human activities represent aggressive, damaging, and unbalanced interactions with wildlife and will increasingly provoke new disease emergencies." They warned that we remain at risk for the foreseeable future.[3]

   In an interview for this book a month after his and Dr. Fauci's research was published, Dr. Morens underscored their findings. "Pandemics are happening more often," he said. An epidemiologist by trade, and current senior scientific advisor at the National Institutes for Health, Dr. Morens has been tracking and studying emerging infectious diseases for over forty years. "The first recorded pandemic was in 430 BC," he said. "And if you plot all the pandemics that we have in historical records since then, they're pretty much episodic. They come every few hundred years. Now, though, we're seeing them come all the time."

Dr. Morens points to repeated influenza outbreaks as an example. "We have had four of them in the last hundred years," he said. "And then we had had Ebola in 2014, and SARS in 2002." In terms of empirical research and statistics, this is, he says, not normal.

Which prompts the question: Why is it happening?

"We know that anybody could say, 'Well, it's just by chance,'" Dr. Morens said. "Maybe there's no way to prove it isn't. But when you think about it, many of these pandemics are associated with human behaviors, and human behaviors exacerbate them."

Dr. Morens is referencing humans' intrusion on wild animals' habitats, which may be connected to the origin of COVID-19. At the start of the pandemic of 2020, the novel coronavirus was attributed to transmission from a bat to a human in Wuhan's wet market, where wild animals, live and dead, are sold.[4]

Wet Markets and Disease

Wet markets are widespread in Asian countries, and many of those countries have large migrant populations. Some experts say the markets are a vital source of livelihood for millions of low-income communities who don't have access to online food-buying options.[5] Live animals like poultry, fish, reptiles, and mammals are slaughtered and sold in these markets for their meat. They're called "wet" because their floors are often hosed down after vendors wash fruits and vegetables or clean fish.[6]

In 1997, live-poultry markets were noted as the source of the H5N1 bird-influenza virus that was transmitted to eighteen people, killing six, in Hong Kong.[7] In 2002, SARS, another virus with roots in human–animal interaction, was first detected in Guangdong, China.[8] According to a 2004 report in the *Lancet*, "The origins of the SARS virus remains unclear, but it is thought that human beings could have contracted the disease by eating or handling civets or other exotic animals from wildlife markets."[9]

Dr. Fauci has called for the global shutdown of wet markets to prevent future outbreaks. "It boggles my mind how, when we have so many diseases that emanate out of that unusual human–animal interface, that we don't just shut it down," Dr. Fauci said in a television interview on *Fox & Friends* in April 2020.

"I don't know what else has to happen to get us to appreciate that . . . because what we're going through right now is a direct result of that."[10]

Dr. Gaurab Basu, an instructor at Harvard Medical School and the co-director of the Center for Health Equity Education & Advocacy, explains that COVID-19 challenges us to think about the wildlife trade and the threat of disease it can bring. "There are dangerous viruses that are coursing through many animals in nature, like bats. They have complicated, nuanced immune systems that can handle certain viruses," Dr. Basu said. He explained that when species like bats remain in their natural habitat, away from humans, the viruses they carry don't pose a threat to people. "The natural environment has figured out how to create ecological equilibrium and stability with the different kinds of viruses that are floating around in there. But once we humans start encroaching into natural habitats, it brings us closer contact with animals that potentially have these viruses. This could cause a lot of disease in human beings," he warned.

Animals Infecting Humans

Scientists estimate that as many as 1.67 million still-unknown viruses infect mammals and birds, and close to 1 million of those may "have the potential to make a *zoonotic leap*—meaning jumping from animals to infecting humans.[11]

Aside from wet markets, Dr. Morens adds that humans may also come in contact with bats, along with other wild animals, through *ecotourism*. These are vacations and tours that allow travelers to visit uneasily accessible natural environments. Increasingly popular in recent years, ecotourism often involves close observation of or interaction with exotic and wild creatures.[12] Protected areas around the globe receive a total of more than 8 billion visits each year.[13]

These international jaunts may evoke images of exotic adventure, but scientists say the trips are ill-advised. Animals' behavior can be altered when their habitats are disrupted. This can impact both the types of pathogens animals carry and the probability of them spreading disease to humans.[14]

Illnesses passed from animals to humans are known as *zoonotic diseases*. According to BBC reports, scientists estimate that three out of every four new or emerging infectious diseases in people come from animals.[15]

That's believed to be the case with Ebola, which killed more than 11,000 people in West Africa between 2013 and 2016. Ebola is transmitted to humans by infected animals, such as chimpanzees, fruit bats, and forest antelope.[16] Bushmeat—non-domesticated forest animals hunted for human consumption—is thought to work as a conduit to transmit the Ebola virus.[17]

Deforestation and Future Pandemics

The quest to hunt and secure food like bushmeat has led to *deforestation*, the permanent removal of trees.[18] The World Health Organization notes that Guinea's forested region, where Ebola emerged, has been as much as 80 percent deforested by clear-cut logging and foreign mining and agriculture companies.[19]

In an interview with ABC Australia, Dr. Christopher Reid said that animals like bats are particularly vulnerable to the effects of deforestation. "It's actually driving bats into areas in greater numbers where perhaps they might not have been before, regions where humans are present," Dr. Reid said.[20]

How fast is the current rate of global deforestation? In 2019, the tropics lost close to thirty soccer fields' worth of trees every single minute.[21] These rapid losses have resulted from many factors, including agricultural expansion, logging, wood harvesting for fuel, infrastructure expansion, and urbanization.[22]

Land development is one of many factors contributing to this problem. Scientists at NASA who interpret forest changes using satellite technology find that rarely is there a single direct cause for deforestation. Most often, they say, it's multiple processes working simultaneously or sequentially.[23]

Deforestation diminishes crucial *biodiversity*—the variety of life in a particular habitat or ecosystem. Tropical forests, for example, contain some of the highest amounts of biodiversity anywhere in the world. But this natural variety is disappearing rapidly as humans clear land to make room for farms and pastures, harvest timber for construction and fuel, and build roads and urban areas.[24]

According to EcoHealth Alliance, deforestation is linked to 31 percent of disease outbreaks such as the Ebola, Zika, and Nipah viruses.[25] *Nature's* online journal *Scientific Reports* found an almost universal two-year link between deforestation and Ebola outbreaks.[26] Their research surmised that areas that experienced significant forest loss were highly likely to see an Ebola outbreak in humans two years later.[27]

Does all of this inevitably point to *more* pandemics?

Dr. Morens believes so. "I don't say this just because as if I have a crystal ball; I say this because of my experience having chased pandemics for forty-five years now. This is the 'New Normal,' and we've got to wake up and deal with it."

Less Biodiversity, More Potential for Disease

Biodiversity encompasses the whole variety of life on Earth. When biodiversity changes occur, the risk of infectious disease exposure goes up—in plants, animals, and humans alike.[28] As reported in the journal *PLoS*, biodiversity loss due to climate change influences the dynamics and distribution of disease-causing pathogens has taken on a new urgency.[29]

Certain animal species are responsible for hosting and proliferating pathogens. "We call them the natural *reservoirs of infection*," said Dr. Rick Ostfeld, senior scientist at the Cary Institute of Ecosystem Studies, a not-for-profit research institution in Millbrook, New York. His research examines ecologically determined risks of exposure to infectious diseases.

Humans are unintentionally raising our risk of getting sick, according to Dr. Ostfeld, through increased greenhouse gases and land development. He says the drive to expand cities and suburbs can harm our health in two ways. The first is through carbon emissions released into the atmosphere from transportation, construction, and manufacturing. Second, by constructing homes and shopping centers and clearing fields to lay asphalt for parking lots, we're destroying the ecosystem that was there before.

Dr. Ostfeld points out that species' reactions vary when their habitat is disrupted. While some perish, others actually thrive, particularly small creatures like rats and fleas. They seem to survive no matter what.

And those, Dr. Ostfeld says, are unfortunately often the ones with the highest propensity for spreading disease to humans.

"They're the rats and mice of the world and the sparrows, even under human disturbance ... urbanization and agricultural land conversion, they do very well, and as a consequence, so do the pathogens they carry," he said.

The bigger animals, though, disappear, and can even go extinct. These are the carnivorous, predatory animals that would eat the tinier disease-infested creatures if left undisturbed, thus reducing their population. Dr. Ostfeld says we are unintentionally eliminating animals that could protect us. "Those larger animals are capable of controlling the abundance of those smaller species," he said.

On a global scale, biodiversity is declining with unprecedented speed, and entire species' extinctions are accelerating.[30] The rate of species loss is now higher than at any time since the dinosaurs went extinct.[31]

Tick Tock: Lyme Disease

Zoonotic diseases are often spread by tiny creatures you may not even notice—that is, until they bite. Pesky ticks, fleas, and mosquitos are not just an annoyance. They can be carriers of pathogens that make you sick. Determining the cause of a patient's infection can often be a race against time for doctors. Patients may initially present symptoms that aren't easily diagnosed, and the ordeal can evolve into a medical mystery.

You've likely heard of two of the most common of these diseases: Lyme disease and West Nile virus. These infections are transmitted by ticks and mosquitoes, respectively. Unfortunately, according to the Centers for Disease Control, the number of cases reported nationally is rising. Instances of mosquito, tick, and flea viruses being transmitted to humans have *tripled* in the US between 2004 to 2016.[32]

So how do weather and climate tie in?

More frequent precipitation and temperature extremes have led to an increase in *vector-borne diseases*: these are illnesses that are transmitted by mosquitoes, ticks, and fleas. Laden with viruses and bacteria, vectors can transfer pathogens from one host to another. Due to climate change, the geographic regions of risk are expanding.[33]

For example, Lyme disease now shows earlier seasonal activity and a generally northward growth. This means that even if Lyme was previously rare where you live, that might not be the case going forward.

Dr. Ostfeld has extensively tracked recent tick migration and its impact on humans. "I think that anyone who's paying attention is surprised by how rapidly Lyme disease is expanding and getting worse," he says. "It's not only spreading geographically: some of that spread is moving upslope up in elevation."

For example, it may have previously been deemed unlikely that you would contract Lyme disease in higher elevations in Canada, but Lyme disease is actually now the most common tick-borne infection in Canada. And it is vastly underdetected. Northern New England, upstate New York, the Upper Midwest, and western Pennsylvania are also locations Dr. Ostfeld cites as similar examples.

The Extended Infectious Season

For vector-borne diseases, the risky season for humans is also getting longer. Climate Central used Stanford University research to analyze the number of days each year when the average temperature is between 61°F and 93°F, the ideal range for vector-borne transmission.[34] They found that of the 244 cities analyzed, almost all (94 percent) have seen an increase in the number of these "disease danger days."[35] This allows ticks and mosquitos more time to come in contact with a host, bite them, and then bite you.

"Ticks seek a mammal or a bird to feed on because that's all they eat, the blood of mammals and birds," said Dr. Ostfeld. "If ticks don't find an animal host, they'll die of starvation. So, if there's a longer period into the fall or early spring for them to find a host . . . then they have a higher probability of surviving."

Climate change is ramping up this cycle of disease-spreading opportunity, and humans are increasingly feeling the consequences. "We have the expansion of Lyme and other tick-borne diseases into new communities where people are unfamiliar with them. That includes both the general population and the health care community. People get exposed but don't really know what's causing their illness," Dr. Ostfeld warns.

The Long Road to Diagnosis

Lyme disease is the most common vector-borne disease in the United States. Symptoms include fever, headache, fatigue, and often a characteristic circular rash. If left untreated, the infection can spread to joints, the heart, and the nervous system.

"With Lyme in particular, as time passes, the bacteria can move into parts of our bodies where they're much more difficult to treat with antibiotics. So expeditious diagnosis and quick treatment is essential," said Dr. Ostfeld.

But that fast diagnosis is not always easy to ascertain, for several reasons. Many of the common symptoms associated with the disease, such as headaches, dizziness, and joint/body pain, are similar to other ailments.

Lyme disease's most distinct symptom—the circular red rash (*erythema migrans*)—does not appear in at least one-quarter of people who are actually infected with Lyme bacteria.[36] And current diagnostic tests do not always detect early Lyme disease since antibodies take time to rise to measurable levels. This can lead to a frustrating experience for both patients and their doctors.

Justin's Journey

Constant headaches, brain fog, cognitive impairment, sensitivity to stimulation, insomnia, joint aches, muscle spasms and tremors, heart palpitations, extreme fatigue——these are all demobilizing symptoms that can be linked to a number of diseases. Connecting this anguish to one tiny tick bite (that often goes unnoticed when it happens) isn't easy. For Canadian Justin Wood, the process took four years and twenty doctors' visits.

Justin, as an outdoorsman in his twenties, spent his summer mountain biking in Alberta. He rode the rough terrain on steep trails and built endurance through twenty-four-hour competitions. Justin enjoyed excursions in the wilderness of northern Ontario.[37] That's where, in 2011, he began to notice something was wrong.

"I was getting increasingly ill with nondescript neurological symptoms," Justin said in an interview with *Explore* magazine.[38] As several years passed, his condition only worsened. It progressed to the point where this previously healthy, athletic young man couldn't participate in sports, school, or work, or even care for himself. His doctors remained puzzled.

Despite not noticing a tick or feeling a bite, Justin thought he should get tested for Lyme disease. He took a test in Canada, which came back negative. But Justin was not convinced. He wanted a second opinion, so he opted for an international test. Sure enough, that one came back positive for Lyme.

Health experts say an initial false negative can happen since antibodies take time to rise to detectable levels.[39] For Justin, receiving the proper diagnosis was a relief in itself.

"I finally had an explanation that made sense," Justin said.[40] After the diagnosis, Justin was able to begin antibiotic treatment. He began to feel better. "It became quite clear Lyme disease was the issue all along," Justin said. "But it took two to three years of treatment before I really started to regain my health."

Justin's passion for the outdoors is rivaled only by his drive to succeed in school, especially science. While battling his illness, the Queen's University graduate completed his master's degree at the University of Calgary, specializing in genetics. How Justin applied that degree to his professional career is, some might say, an example of how everything happens for a reason. Justin's arduous journey to a diagnosis, along with his background in genetics, inspired him to help others in similar circumstances. He created his own company to improve Lyme disease testing. "I've always loved biology. After going through my journey with Lyme disease, I thought it made sense to work in a field that allowed me to help with this massive problem."[41]

Justin launched Geneticks, the first private lab in Canada solely dedicated to providing testing ticks for tick-borne diseases.[42] Geneticks's rapid-return service includes information about the presence of tick-borne pathogens in submitted ticks within two to five business days of arriving at their laboratory.[43]

"One need that I noticed how important it was to know quickly if you'd been exposed, and then decide the next steps with your health-care provider," Justin said, recalling his five-year ordeal to canadiantraveller.com, adding: "Lyme disease can cost you so much: your health, finances, and quality of life."[44]

Tick Disease Discoveries

For doctors, solving the puzzle of matching symptoms to the cause is not clear-cut—particularly if the vector-borne disease has yet to be discovered. Many have heard of Lyme disease, for example, but new diseases are still emerging.

Dr. Dana Hawkins, MD, an infectious disease specialist at the University of Kansas Hospital in Kansas City, KS, is a pioneer in this research. "I helped discover *Bourbon virus* a few years ago, and that's a tick-borne illness," Dr. Hawkins said, explaining the unintentional outcome.

"I was looking for a different virus originally—the *Heartland virus*. But I knew that this patient just was not reacting the way that somebody should if they had Rocky Mountain spotted fever, even though they were getting the right antibiotic, so we just had to dig a little bit deeper. With help from the CDC, we were able to identify this new virus now," he said of his 2018 discovery.

Rocky Mountain spotted fever has been around a long time, so doctors are more familiar with its signs and symptoms than other tick-borne diseases. It was first noted in 1896 in the Snake River Valley of Idaho. Initially called black measles because its related rash turned the skin black, the disease often proved fatal. It didn't take long for the spread to occur. By the early 1900s, cases were reported in Washington, Montana, California, Arizona, and New Mexico.[45]

Heartland virus was discovered in Missouri over a century later, in 2009. It can be transmitted not only by ticks but also by mosquitos and sandflies. As for the Bourbon virus recently discovered by Dr. Hawkins, it's been identified in a limited number of patients in the Midwest and the southern United States. Currently, CDC scientists "do not know if the Bourbon virus might be found in other areas of the United States."

How did the Bourbon virus get its name? "You know," Dr. Hawkins recalled, "I wanted a different name, but the CDC based the name on where they did the lab work to discover this new isolate, and named it after the county in Kansas that the patient was from"—and that was Bourbon County.

The Wrath of the Mosquito

"The mosquito has ruled the earth for 190 million years
and has killed with unremitting potency for
most of her unrivaled reign of terror."[46]

Author Timothy C. Winegard writes in his 2019 book, *The Mosquito: A Human History of Our Deadliest Predator*, that this "nefarious pest, roughly the size and weight of the grape seed, is estimated to have taken the lives of 52 billion people throughout history."[47]

What to Do if You Find a Tick on Your Body

"Be aware of what ticks look like and remove them quickly if you find them," advises Dr. Ostfeld. Here are more of his expert tips:

- USE a pair of fine-tip tweezers.

- GRAB the tick as close to the skin as possible.

- PULL straight out. Ticks secrete a chemical that's very much like cement, so they anchor themselves pretty firmly in your skin. You can pull them out, but sometimes their mouth or other body parts will remain in the skin.

- DAB the area of the bite with some alcohol, then leave it alone.

- DON'T use nail polish or a hot match or petroleum jelly. These don't work, and they are slowing down the process of tick removal. The faster you pull out the tick, the less likely it is to transmit whatever pathogens it's carrying.

- DON'T waste time. If you don't have tweezers, use your fingernails or fingertips.

- SAVE the tick if possible. Identifying it might be helpful to your health-care provider. Acting quickly and calmly is of the essence.

Clearly, the blood-thirsty mosquito is nothing to make light of, yesterday or today. One concern is that it can carry dangerous diseases like *West Nile virus* (WNV), which is the leading cause of mosquito-borne disease in the continental United States.[48] Around one in five people who are infected with WNV develop a fever and other symptoms. For roughly one in 150, WNV can grow into a serious, sometimes fatal, illness.[49]

According to Climate Central, mosquito-friendly weather occurs when temperatures are between 50°F and 95°F, with relative humidity above 42 percent. Mosquito mortality rate goes up outside these ranges.[50] Traditionally, cases of WNV occur during mosquito season, which starts in the summer and continues through fall. But as with tick season, due to our changing climate, mosquito season is getting longer, giving the bugs more time to thrive.

Increasingly warmer temperatures can even accelerate their quest for human blood. "They might bite two or three times more frequently when it's several degrees warmer, on average," Dr. Ostfeld said.

Here are the gruesome details: after female mosquitoes ingest blood from infected birds, the virus replicates within the gut of the mosquito and its salivary glands. It is transmitted in the salivary fluid to humans or birds during subsequent bites.[51]

Where do these biters like to hang out? Mosquitoes live in cities and suburbs, according to Dr. Ostfeld. "They aren't really rural."

Increased humidity and more frequent heavy rain events have also helped to expand the ideal conditions for mosquitoes to reproduce and potentially infect humans.[52] The National Climate Assessment studied the Asian tiger mosquito, which can transmit West Nile and other diseases.[53] For the northeast United States, scientists project there will be three times as much suitable land for some types of mosquitoes to inhabit by the year 2035.[54] Science author Carl Zimmer explains in an essay from his book *A Planet of Viruses* that hot and humid conditions can speed up the growth of the viruses mosquitoes are carrying.[55]

Epidemiologist and medical historian Dr. David Morens adds that as the temperature rises, outdoor vegetation and foliage thickness also increases, which mosquitoes take advantage of. "This gives the mosquitoes more habitat to hide in, and it also changes the behavior patterns of birds," he said. "Some birds eat mosquitoes, but mosquitoes also bite birds, and some of these infections, like West Nile, are regulated by mosquitoes biting birds."

The Discovery of West Nile Virus

In New York City in 1999, several people became deathly ill with encephalitis—an inflammation of the brain, most commonly caused by a viral infection. At the same time, and in the same area, crows were reportedly dying in large numbers.

Were these mysterious deaths related?

An epidemiologist at the city health department theorized that the large numbers of dead birds might be connected to human cases of encephalitis. The CDC analyzed blood samples from the people who died, and initial results suggested that the deaths were the result of St. Louis encephalitis (SLE), a disease that had previously occurred in the area and that can be transmitted from birds to humans by mosquitoes.

Bronx Zoo head pathologist Dr. Tracey McNamara investigated the cause behind the increase in crow illness and death. She noted that it wasn't only crows that were dying—several of the zoo's flamingos and one of its bald eagles also died. Were their deaths connected?

An analysis of samples taken from the dead zoo birds performed by the US Department of Agriculture National Veterinary Services Lab in Ames, Iowa, revealed that the deaths weren't caused by an SLE virus but by something else. It soon became apparent that the same virus was causing both bird and human deaths, and that this was a newly emerging disease.

Nearly three months after the initial outbreak, government scientists announced that the culprit was the West Nile virus, which had never previously been found in the Western Hemisphere.

Since 1999, there have been 1.5 million West Nile infections in the United States.

(Source: OneHealthCommission.org)[56]

West Nile Virus Expansion

Like Lyme disease, West Nile virus can be difficult to diagnose, especially when it occurs in unlikely locations. Dr. Gaurab Basu, physician and instructor at Harvard Medical School and the founding co-director of the Center for Health Equity Education, shares his story of an unexpected—and scary—diagnosis in Massachusetts:

> I had a patient who was doing well. He was around seventy years old. I got an alert on my email the way I do when one of my patients gets admitted to the hospital. He seemed quite sick. He had a fever. He was very confused, disoriented, and having a difficult time talking. It took a long time for the hospital team to try to sort out what was going on because those are very nonspecific complaints.
>
> The first thing we do is a lumbar puncture to take some fluid around the spine to evaluate if there's an infection there. What happened after those studies came back ... well, the results were kind of a stunning diagnosis: *West Nile Encephalitis*. It was completely unexpected. I think one of my reactions was, "My gosh." I don't think about West Nile when I think about patients with the symptoms that he came in with. But in retrospect, it all made sense.
>
> We call it a *differential diagnosis* in Massachusetts. If I was working in areas closer to the equator, certainly that's much more on people's minds. There you have a number of mosquito-borne illnesses like dengue and malaria and things like that. But up here in Massachusetts, it's not something we think about. It's a very serious disease. Not only was the patient infected with this virus, West Nile, but it had gotten into the fluid around his spine and brain. Suddenly, we could explain why he had been so disoriented and why his fever had increased over the hospitalization. He went from being a little bit confused to be completely disoriented, unable to put sentences together, not knowing who he was or where he was.

Dr. Basu says the patient survived, but West Nile encephalitis took a severe toll on his health. He had trouble reading and concentrating and relied on his partner for support. This lasted for many months, almost a year.

The WNV-infected man had later recalled that before becoming ill, while he was outside, a mosquito bit his neck. "For me as a doctor," said Dr. Basu, "I reflect on how health and disease is not just a patient and their body, their anatomy, and physiology, but it's also how they interact with the world and the environment."

Zika Virus and Mosquito Inequity

First identified in Uganda in 1947 in monkeys, then humans in 1952, the *Zika* virus is transmitted through the bite of an infected mosquito. The first large outbreak of Zika wasn't reported until 2007, on the island of Yap in Micronesia.[57]

Later in 2015, Brazilian investigators reported that Guillain-Barré syndrome (GBS), a rare neurological autoimmune disorder, which had been previously recognized among some patients with Zika virus disease, was also associated with virus infection during pregnancy. This could result in congenital *microcephaly*, a congenital disability where a baby's head is smaller and/or the head stops growing after birth.

The World Health Organization declared the Zika virus, related microcephaly clusters, and other neurologic disorders a "Public Health Emergency of International Concern" in 2016.[58]

Later that year, Zika virus transmission had expanded to forty-eight countries and territories, including North and South America.

The demand for professional pest control to deter disease-spreading mosquitoes skyrocketed during the Zika outbreak. According to the "State of the Mosquito Market Report," mosquito control revenue increased from 2015 to 2016 by 30 percent.[59]

In the article for otherwords.org, "The Mosquito Gap: How Poverty, Climate Change, and Bad Policy Put Poor People at Greater Risk from Pest-borne Diseases," author Sarah Anderson explains that for low-income Americans, professional pest control services that feature automatic home spraying systems can be cost-prohibitive. Yet, unfortunately, it's the poorest neighborhoods that need these services the most.[60]

In one study in Baltimore, scientists evaluated Asian tiger mosquito populations across various neighborhoods, ranging from low to high income. Over three years, they noted where mosquitoes tended to breed and bite. Their favored spots? Abandoned houses and buildings, around accumulating trash, in standing water, and on surface vegetation—all sites mainly found in the lower-income sections of Baltimore.[61] These scientists urged that socioeconomic factors need to be considered when modeling mosquitoes' threats and their potential for disease spread.[62]

Another study in Chatham County, Georgia, found that indigent and minority populations in urban areas are often at higher risk of infectious diseases like the West Nile virus than other demographic groups.[63] Health scientists from Liberty University released their 2017 study, which gathered and evaluated data for a decade in Savannah. They identified people in the areas they found most at risk. They included:

- Elderly
- Women
- Minority and disadvantaged communities (4.5 and 5.5 times more likely to be at risk)

One of the reasons the study gave for their findings was that the wealthier neighborhoods "were nineteen times more likely to contact and receive services from the Chatham County Mosquito Control Department."[64]

Outbreaks of the Zika virus in Brazil and Puerto Rico were also mapped by various studies connecting the cases with impoverished areas. Public health scientists at Boston University point out that "the rates of mosquito-borne disease transmission are dramatically different between cities that have well-developed systems of control for stagnant water and those that do not."[65]

Dr. Cheryl Holder, MD, has personally seen these inequities affect her patients in Miami, Florida. Dr. Holder is a key medical provider dedicated to serving underprivileged populations. Many of her patients live in older, poorly constructed, heat-trapping apartments or homes with no screens on their windows.

"There's no legislation or regulations that say you need screens on the houses. If you're a renter and you can't afford the AC, you can't close all your windows, and you don't have screens. This makes you much more vulnerable to Zika," Dr. Holder said.

Plus, heavy rainstorms are more common during Florida's wet season, which means more standing water opportunities. "There's more water pooling for mosquito breeding. They bite day and night," she said.

According to the CDC, Zika can be passed from a pregnant woman to her fetus. Infection during pregnancy can cause certain birth defects.

Dr. Holder previously treated a pregnant South Florida woman and contracted Zika through a mosquito bite she received while working outdoors on a farm. Her patient was relieved when she learned that, fortunately, the virus was not passed on to her baby. But in the months and even year that followed the baby's birth, the patient had postpartum depression from all the stress.

How to Reduce Your Risk For Mosquitos

Insect repellent: when used as directed, Environmental Protection Agency (EPA)–registered insect repellents are proven safe and effective, even for pregnant and breastfeeding women. Use an EPA–registered insect repellent external icon with one of the following active ingredients:

- DEET
- Picaridin
- IR3535
- Oil of lemon eucalyptus (OLE)
- Para-menthane-diol (PMD)
- 2-undecanone

The EPA's search tool is available at https://www.epa.gov.

- Always follow the product label instructions.

- Reapply insect repellent every few hours.

- Do not spray repellent on the skin under clothing.

- If you are also using sunscreen, apply sunscreen first and insect repellent second.

- Use an EPA-registered insect repellent.

- Follow instructions when applying insect repellent to children.

- Do not use insect repellent on babies younger than two months of age.

- Dress your child in clothing covering arms and legs, cover crib, stroller, and baby carrier with mosquito netting.

- Treat clothing/gear with permethrin or purchase permethrin-treated clothing and gear.

- Do not use permethrin products directly on the skin.

- Use screens on home windows and doors. Repair holes in screens.

- Use air-conditioning when available.

- Once a week, empty and scrub, turn over, cover, or throw out items that hold water, such as tires, buckets, planters, toys, pools, birdbaths, flowerpots, or trash containers.

(Source: CDC)[66]

Are Mosquitoes Attracted to Some People More Than Others?

One thing all mosquitoes have in common is their complex sense of smell.[67] Mosquitoes must seek out humans to bite because our blood provides the nutrients they need to reproduce (only the females bite). Interestingly, these hungry insects have poor vision, so they use their keen sense of smell to seek their meals. Biologically, mosquitoes have "exquisitely sensitive small hairs, called sensilla, on their antennae and mouthparts."[68] These hairs have scent receptors that help them select who to bite.

Researchers have long known that mosquitoes are attracted to the scent of human sweat,[69] which includes the odor of lactic acid. Bacteria emitted from your skin, mixed with bacteria on the skin's surface, contribute to your unique scent.

When it comes to which person to bite, Dr. Ostfeld explains, mosquitoes favor "people who have less efficient metabolisms, so they produce more heat, more carbon dioxide than another person would. That's more attractive to mosquitoes." If you're exhaling more significant volumes of carbon dioxide, that may make you an easier target for mosquitoes.[70]

While some people attract more mosquito bites than others, it's unlikely that someone would never, ever be bitten.[71] Health experts advise even those who say "Mosquitoes never bite me!" to not be complacent. They should still take precautions. It only takes one bite to contract a mosquito-borne disease.

Flesh-Eating Bacteria

In the summer of 2019, Cheryl Bennett Wiygul's parents drove from Tennessee to Florida for a weekend visit to spend time with her and their grandchildren. In a public Facebook post, she describes how what started as a fun day quickly took a tragic turn.[72]

> We were out in the bay on the boat near Crab Island, went to the beach in Destin . . . Daddy stayed up late Friday night and watched a movie. He was happy and talkative—about 4:00 a.m. Saturday morning, twelve hours after we were in the water, he woke up with a fever, chills, and some cramping.
>
> He got worse (later that morning). His legs started to hurt severely. He was becoming extremely uncomfortable. My dad is not a complainer, so he had to have been in a lot of pain to vocalize it. . . . As they helped him get changed into his hospital gown, they saw this terribly swollen black spot on his back that was not there before. My mom sent me a picture of it. . . . The black spot had doubled in size. A new one was starting to pop up. His arms were becoming blotchier by the minute. He was in a great deal of pain. Some of the nurses said they'd never seen anything like it.
>
> At 1 a.m. (the following day), he became septic, and they moved him into the ICU. They said his organs were too damaged, and his blood was too acidic to sustain life.
>
> He was gone by Sunday afternoon. Less than forty-eight hours after getting out of the water feeling great, the bacteria had destroyed him.[73]

Cheryl said that lab reports showed her father had "*vibrio vulnificus*, which manifests into necrotizing fasciitis (flesh-eating bacteria), ultimately leading to sepsis."[74]

"Flesh-eating bacteria sounds like an urban legend," Cheryl wrote. "Let me assure you that it is not. It took my dad's life . . . There isn't enough education out there about the bacteria in the water."[75]

Bacteria, common single-celled organisms, are found naturally in lakes, rivers, and streams. Most are harmless to humans; however, certain bacteria, like the Vibrio species, can infect people with a disease known as vibriosis. The CDC estimates that vibriosis accounts for eighty thousand illnesses each year in the United States.[76]

Existing in certain coastal waters, these bacteria are present in higher concentrations between May and October, when water temperatures are warmer. *Vibrio vulnificus* infections are primarily transmitted in two ways. The first is through raw or undercooked seafood (they are the leading cause of death related to seafood consumption in the US).[77] The second is through contact when swimming in contaminated salt brackish waters (a mixture of fresh and salt water). The infection enters through an open wound on the body.

Those who become infected with *Vibrio vulnificus* may need intensive care or limb amputations. About one in five people with this infection die, sometimes within a day or two of becoming ill.

Cheryl and her mother did not notice any open wounds on her dad before they went swimming. She later learned that her father was at greater risk for this infection because he had a compromised immune system due to a previous illness. "I feel like I should have known, and that is something I will live with for the rest of my life. If I had done more research, I would have, but I don't think the general public realizes it either," she wrote in her Facebook post.[78]

Those with preexisting health conditions like liver disease, diabetes, kidney failure, cancer, or a compromised immune system are at greater risk for *Vibrio* illness. The CDC says severe complications are also more likely for those with thalassemia or HIV and those who receive immune-suppressing therapy for disease treatment, take medicine to decrease stomach acid levels or have had recent stomach surgery.[79]

Vibrio Rising

Recent data have shown that the incidence of *Vibrio*-associated illnesses is increasing worldwide, up by 41 percent between 1996 and 2005.[80] In more recent years, the number of infections associated with swimming in coastal waters has grown, correlating with record summer heat waves, like the one in 2016.[81]

Dr. Geoff Scott, PhD, the chair and a clinical professor at the Environmental Health Sciences Arnold School of Public Health at the University of South Carolina and a leading toxicologist in Charleston, has been tracking *Vibrio* abundance for decades. "What does climate change have to do with it?" Dr. Scott asks. "By extending the temperatures for a longer period to more places, that's making *Vibrio* more prominent. The illness used to run prime early in August and September. Now we're seeing illnesses from April to the end of October because temperatures are warmer. We've seen it in the Gulf of Mexico up to the Gulf of Alaska and North Sea Coast of Britain."

Dr. Scott also pointed out that hotter temperatures can impact the body's natural ability to fight infection. "I think the virulence of the bacteria [the ability of an infectious agent to cause disease] is getting more [rapid and severe] and antibody resistant. I believe that *Vibrio*s are a real bellwether for climate change."

Dr. Scott is working with a team of researchers to establish a *Vibrio* forecasting system.[82] "What we're now trying to do is to move to the second generation of models, where we don't just predict the abundance of *Vibrio*s but also the virulence and [levels of] antibiotic resistance," says Dr. Scott.

Reduce Your Risk of Vibriosis by Following These Tips

- Don't eat raw or undercooked oysters or other shellfish. Cook them before eating.

- Always wash your hands with soap and water after handling raw shellfish.

- Avoid contaminating cooked shellfish with raw shellfish and its juices.

- Stay out of saltwater or brackish water if you have a wound (including from recent surgery, piercing, or tattooing), or cover your wound with a waterproof bandage if there's a possibility it could come into contact with saltwater or brackish water, raw seafood, or raw seafood juices. (Note also the other health factors that increase a person's risk of vibrio infections discussed earlier in this chapter.)

- Brackish water is a mixture of fresh and saltwater. It is often found where rivers meet the sea.

- Wash wounds and cuts thoroughly with soap and water if they have been exposed to seawater or raw seafood or its juices.

- If you develop a skin infection, tell your medical provider if your skin has come into contact with saltwater or brackish water, raw seafood, or raw seafood juices.

- Wear clothes and shoes that can protect you from cuts and scrapes when in saltwater or brackish water.

- Wear protective gloves when handling raw seafood.

(Source: CDC)[83]

More Ways to Reduce Your Risk of Lyme, West Nile, and Zika

Before You Go Outdoors

- Know where to expect ticks: grassy, brushy, or wooded areas, or even on animals. Spending time outside walking your dog, camping, gardening, or hunting could bring you in close contact with ticks. Many people get ticks in their own yard or neighborhood.

- Treat clothing and gear with products containing 0.5 percent permethrin, which can be used to treat boots, clothing, and camping gear and remain protective through several washings. Alternatively, you can buy permethrin-treated clothing and gear.

- Use EPA-registered insect repellents containing DEET.

- Follow product instructions. Do not use products containing OLE or PMD on children under three years old.

- Walk in the center of trails.

After You Come Indoors

- Check your clothing for ticks. Any ticks that are found should be removed (see page 65 for instructions).

- Tumble dry clothes in a dryer on high heat for ten minutes to kill ticks on dry clothing after you come indoors. If the clothes are damp, additional time may be needed.

- If the clothes require washing first, hot water is recommended. Cold and medium temperature water will not kill ticks.

- Examine gear and pets. Ticks can get into homes, then attach to a person later, so carefully examine pets, coats, and backpacks.

- Shower soon after being outdoors. Showering within two hours of coming indoors has been shown to reduce your risk of getting Lyme disease and possibly other tick-borne conditions.

- Conduct a full-body check upon return from potentially tick-infested areas, including your backyard.

- Use a hand-held or full-length mirror to check these parts of your body and your child's body for ticks:
  - Under the arms
  - In and around the ears
  - Inside belly button
  - Back of the knees
  - In and around the hair
  - Between the legs
  - Around the waist

(Source: CDC)[84]

5.

THE NEW
ALLERGY SEASON

Does allergy season seem like it's getting longer?

If you are one of the 25 million Americans who find it harder to breathe in the spring due to allergies, that fateful first sneeze is indeed arriving sooner than it used to. Allergy season is now estimated to be eleven to twenty-seven days longer than it was just two decades ago.[1]

Pollen, one of the most common triggers for allergies, may also be getting stronger. Warmer temperatures and projected levels of increased carbon dioxide in the atmosphere, according to researchers, enable plants to potentially produce more allergenic pollen in larger quantities. This means that many locations could experience extended allergy seasons *and* higher pollen counts as a result of climate change.[2]

Plant-produced pollen isn't the only factor aggravating allergy sufferers. Air pollution, from ozone gas to wildfire smoke, is especially harmful to those with allergies and asthma. Smog, for example, can aggravate the allergenicity of certain pollens, and disperse tiny particulate matter that's more readily inhaled.

Other outdoor conditions can also be sneeze-inducing in unexpected ways. Severe weather events like thunderstorms can trigger asthma attacks. Light rain showers, though, can have a mitigating effect and reduce pollen spread.

Some are more susceptible to these hidden dangers than others. Children and minorities living in low-income communities, for instance, face a greater likelihood of developing asthma. But doctors say there are precautions you can take to reduce the risk of allergen interaction both outside and inside your home.

Nose Nemesis

Nature's bounty in bloom is beautiful but nondiscriminatory. The notorious nose nemesis known as ragweed blossoms each season. Soft-stemmed and filled with billions of lightweight pollen grains that float easily through the air, ragweed produces particles that can travel for many miles before they find themselves irritating your nose and throat. An estimated 15.5 percent of all Americans are sensitive to ragweed.[3]

How does exposure to ragweed kick your upper respiratory system into overdrive? It starts with an *allergy*, which occurs when your immune system sees a substance as harmful and overreacts to it. The substances that cause these allergic reactions are *allergens*, which can be inhaled, ingested, or absorbed through the skin.[4]

The biological response to an allergen can happen quickly. It begins with an increase in blood flow and inflammation, narrowing your airways, making it harder to breathe. Then your brain signals a sneeze to try to expel the pollen. Membranes in your nose start to make more mucus, leaving you with a runny or stuffy nose. The mucus also runs down your throat and causes you to cough. Histamines, which your immune system uses to protect you, can also make your eyes and nose itch.

When someone has allergies, their immune system makes antibodies to attack the allergen. These antibodies are particular—a person can be allergic to one type of pollen, for example, and not another. It's possible that, upon repeated exposure, the severity of the reaction may increase.[5]

First Bloom

Timing is everything in nature. Notable signs of spring's imminent arrival can be a sprig of leaves appearing on a formerly bare tree branch or crocuses pop-

ping up through the surface of snow-covered grass. Yet this natural phenologi-
cal timing is no longer occurring like clockwork; its innate calendar is shifting
due to climate change.

First-bloom indicators that spring has sprung are showing up sooner.[6]
Over the past thirty years, Climate Central researchers found that *"leaf out"*
(when trees sprout leaves) is happening earlier in 76 percent of the US cities
they analyzed.[7] Colorado Springs, for example, has seen leaf-outs occur around
seventeen days sooner than it did thirty years ago.[8]

Pollen's Late Check-Out Time

Allergy season isn't just starting ahead of schedule; it's ending later as well. Uncomfortable symptoms like watery eyes and sinus congestion usually dissipate
with the first frost in the fall. But that date has shifted, depending on where
you live.

Climate data compiled from the National Oceanic and Atmospheric
Administration's National Centers for Environmental Information (NOAA)
shows many US states have experienced the first frost of fall several days later
than they did, on average, over a century ago—in some cases, over two weeks
later. For example, in Arizona, the first frost of fall comes about twenty days later
than it used to; Michigan is about six days later, and Florida around twelve.[9]

These early springs and late falls add up to an extended *frost-free season*.
This refers to the stretch of time when temperatures remain above 32 degrees.[10] Witness pollen's extended stay in your sinuses.

Logistically, the longer allergy season generally becomes more pronounced
from south to north, according to data analyzed from the years 1895 to 2015.
For example, data shows that ragweed season increased by twenty-five days in
Winnipeg, Manitoba; by twenty-one days in Fargo, North Dakota; by eighteen
days in Minneapolis; and by six days in Oklahoma City.[11]

This extended stretch is also of consequence to asthma sufferers, as allergies can trigger attacks. Emergency-room visits for a severe asthma attack are
scary, but unfortunately, they occur often. In 2015, the CDC reported 1.7 million ER visits with asthma as the primary diagnosis. Asthma is a leading cause

of child emergency-room visits and hospitalizations, and missed school days.[12] Annually, thirty-five thousand to sixty thousand ER visits for asthma are linked with exposure to oak, birch, and grass pollen, according to a study published in January 2019.[13]

Climate scientists warn that if emissions of heat-trapping gases continue to grow, emergency departments could see an additional 3,700 cases each year by 2030. That number could swell to another ten thousand annually in 2090.[14]

Emboldened Mold

If you're sneezing year-round, your allergy may not be seasonal. Mold spores or other types of fungi could be the culprit. Molds are microscopic organisms that live everywhere. Their seeds, called *spores*, travel through the air. Some spores spread in dry, windy weather. Others circulate with the fog or dew when humidity is high. Mold in the air outside can also attach itself to clothing, shoes, and pets, all of which can carry mold indoors. When mold spores drop on spots where there is excessive moisture, such as where leakage may already have occurred in roofs, pipes, walls, and even plant pots or places where flooding has occurred, mold will grow.[15]

Rotting logs, fallen leaves, compost piles, grasses, and grains are additional mold breeding grounds. At home, you may encounter mold growing in bathrooms, kitchens, or baasements.[16] Damp conditions increase the likelihood of occupants' exposure and resulting health effects. For those who are allergic, inhaling mold spores can trigger an asthma attack.[17] Spore allergies are most common from July to early fall. But fungi growth can exist both inside and outside your home, so allergic reactions can occur year-round.

Although there are numerous mold types, only a few dozen cause allergic reactions. Sometimes it can be hard to identify which one is causing you the most problems. According to the Mayo Clinic, being allergic to one type of mold doesn't necessarily mean you'll be allergic to another. Some of the most common molds that cause allergies include alternaria, aspergillus, cladosporium, and penicillium.[18]

Exposure to indoor fungi and mold may trigger symptoms similar to hay fever, including:

- Headaches
- Sneezing
- Runny nose
- Red eyes
- Skin rashes

According to allergists, more frequent flooding events and rising sea levels could have a worsening effect on asthma because of the potential increase of mold in both indoor and outdoor air.[19] These climate-related events may break down the physical barriers between outdoor and indoor spaces. This deterioration could allow more water and moisture to permeate the home, particularly in areas that are already prone to being leaky. These spots may face exacerbated mold problems.[20]

Pollen Power

Allergy season isn't just longer; it's also getting stronger. Scientists believe that rising levels of CO_2 in the atmosphere due to climate change may have a supercharging effect on specific plants and their pollen.[21]

Dr. Lewis Ziska is a former researcher with the United States Department of Agriculture (USDA), and is now an associate professor of environmental health sciences at the Columbia University Irving Medical Center in New York. His research shows that, in addition to the lengthening allergy season, increased CO_2 by itself can elevate the production of plant-based allergens.

"You often hear that phrase 'CO_2 is plant food.' It's not just food for plants that you like; it's plant food for everything, including those plants that generate pollen," Dr. Ziska said. "So, all of these things are in combination. We think, based on the evidence that we have, that we are changing the allergy seasons, making it longer and increasing the amount of pollen that you come in contact with."

But will rising CO_2 levels intensify the *allergenicity* or allergic potential of pollen?

"We're still debating about whether it's more potent—that is, whether the same amount of pollen will induce *more* allergies," Dr. Ziska explained.

Dr. Ziska points out that two characteristics have decidedly changed based on evidentiary data: allergy season length and the resulting additional pollen accumulated over time. This means there's more time each year for those with pollen allergies to experience sneezing, runny nose, and itchy eyes.

Israeli scientists believe rising CO_2 levels will cause particular fungi that emit mold spores to grow more allergenic. They found that mold grown in current carbon dioxide levels produced 8.5 times as much allergenic protein as those grown in pre-industrial carbon dioxide levels.[22] Their hypothesis is that carbon dioxide–induced changes amplify the bacteria's respiration and growth process.

Another study, led by William Anderegg of the University of Utah's School of Biological Sciences, found that human-caused climate change played a significant role in allergy season lengthening and contributed to increased pollen amounts. The team compiled measurements between 1990 and 2018 from sixty pollen count stations maintained by the National Allergy Bureau across the United States and Canada. These stations collected airborne pollen and mold samples, which were then hand-counted.[23]

Applying statistical analysis in conjunction with nearly two dozen climate models, the researchers found that nationwide pollen amounts increased by around 21 percent over the study period, with the greatest increases recorded in Texas and the midwestern United States.

The results showed that climate change alone could account for around half of the pollen season lengthening and around 8 percent of the pollen amount increasing. "This study reveals that connection at continental scales, and explicitly links pollen trends to human-caused climate change," said Dr. Anderegg.[24]

Predicting Pollen

If you live in the South, finding a thick coat of yellow dust accumulating on your car's windshield is a telltale sign that pollen will be a problem that day. The more exact science of accessing daily pollen counts involves tabulating the number of tiny grains found in one cubic meter of the air. The higher the count, the greater the chance that those suffering from hay fever will experience symptoms outdoors or exposed to the outside air.[25]

There is a difference, though, between a *pollen count* and a *pollen forecast*. Pollen counts are measured by the National Allergy Bureau from observation stations in real time.[26] Meteorologists and scientists predict *future* pollen levels using those counts, plus incorporating past seasonal pollen information and predictive weather models. The pollen forecast is a common feature on TV weathercasts in the spring!

There are interesting connections between daily weather and its influence on the potential impact of pollen particles. Rain, for example, can have a cleansing effect on pollen in the air. Light, steady showers can wash the pollen away and prevent it from flying about.[27] Here's how it works: the tiny drops of a light rain shower absorb pollen particles through the process of *coagulation*. As raindrops fall, they develop a small electric charge that attracts particles in the air. Small drops actually have more surface area per volume than large drops. As reported in sciencing.com, "This electric charge and the larger surface area combine for more coagulation and a better cleaning."[28]

Wind can play an important role as well. Spring breezes fertilize flowers, a natural process that can spread allergens farther than you might think. The website AllergicLiving.com cites some impressive pollen-travel distances. For instance, a University of Tulsa aerobiologist detected cedar pollen in Oklahoma that had flown over a thousand miles from Ontario, Canada. It took about a day and a half to get there.[29]

Thunderstorm Asthma

When you see lightning or hear thunder, it's a weather warning indicating thunderstorms are nearby. Retreating indoors for safety is advisable. But for those with allergies and asthma, stormy weather can bring additional danger, even after the storm has retreated. Thunderstorms can trigger allergies and even severe asthma attacks.

This might seem counterintuitive. If light rain showers can slow down or stop pollen, then wouldn't the heavy rain and strong winds in a thunderstorm make that coagulation process happen even faster?

Let's first look at the meteorology. In the early stage of a developing storm, the towering cumulus stage, clouds grow vertically. A rising updraft of warmer air will suspend raindrops until the point where their weight is too heavy. That's when a strong downdraft of colder air, along with falling rain, descends forcefully from the cloud. A "*severe thunderstorm*" is classified as having winds of 39 mph or greater. Even when thunderstorms aren't severe, they can contain winds that can blow in any direction. This is why rain can appear to fall sideways sometimes.

That rapid up/down motion within storms is strong enough to rupture the exterior coating on pollen's tiny grains, releasing even smaller allergy-inducing fragments. These are nearly a tenth of the original pollen grain size, and are so lightweight that they don't settle readily to the ground. They often remain aloft for hours, lingering in the air. These invisible, microscopic allergens are inhaled much more deeply into human airways than full-size pollen grains. The thunderstorm-induced allergic response can result in serious health complications in patients with asthma or hay fever.[30]

The medical data matches up with the meteorology. Doctors report more ER visits for asthma when thunderstorms appear in the forecast. According to a research letter in the *Journal of the American Medical Association* (*JAMA*) in August 2020, thunderstorms are associated with an average of more than three thousand additional emergency-room visits annually among older adults with asthma and chronic obstructive pulmonary disease (COPD) in the United States.[31]

Climate research indicates that rising sea surface temperatures are increasing the frequency of thunderstorms and fueling greater intensity in tropical storms.[32]

An Ashy Trigger

About half of the world's air pollution comes from wildfires. The year 2020 was the most active wildfire year ever across the American West, with five of the six largest fires in California history and the three largest fires on record in Colorado.[33] Data shows that the number of large wildfires in the West has more than doubled since the 1980s and expanded their reach into previously unaffected areas, all triggered by climate change.[34]

Warmer temperatures causing earlier snowmelt and longer-lasting droughts are contributing to more frequent fires. Dry vegetation and low soil air moisture make it easier for fires to spread quickly. Under these conditions, a spark from a lightning strike, electrical failure, human error, or planned fires can quickly get out of control.[35]

Wildfire smoke contains a mix of gases and fine particles from burning vegetation, building materials, and other elements, releasing large amounts of carbon dioxide, black carbon, brown carbon, and ozone into the atmosphere. This smoke can remain in the air for weeks and travel thousands of miles. Wildfire pollution impacts the health of even those who don't live near them.

For example, smoke from fires burning as far away as Siberia can impact those living in British Columbia. In 2012, NASA scientists tracked smoke from large wildfires in Siberia that soared high enough into the atmosphere for winds to push the plumes across the Pacific Ocean to North America. The result was dangerous air quality for Canadians, with record highs of ground-level ozone.[36]

Smoke from wildfires emits tiny particulate matter that is difficult for anyone to breathe. But those with asthma and allergies are even more susceptible to health problems. Allergist-immunologist Dr. Suzanne Cassel of Cedars-Sinai Hospital in Los Angeles explains: "People with allergies and asthma often have chronically inflamed airways, which makes them very sensitive to irritants like smoke. They may experience worsening of their allergic symptoms or develop an asthma attack even to low levels of environmental smoke."[37]

While strong winds in the upper levels of the atmosphere transport wildfire smoke, burning wildfires can also influence the weather at the ground level—causing some interesting phenomena. Large fires can alter local weather patterns. In extreme conditions, these wildfires can create their own weather in the form of *firestorms* and *firenadoes*.

Envision swirling winds of flames rising into the sky. As heat is constantly and quickly rising from the fire, the air is forced upward. The surrounding air rushes in to fill that gap. That movement is so intense and rapid; it creates a mighty wind called an *updraft*. In some cases, the rising air whirls up so fast it creates a fire tornado.[38]

As more fires burn, they also release increased amounts of pollutants into the atmosphere. This can trigger hotter and drier weather and, thus, more fires. They also destroy crucial trees that remove CO_2 from the air—a dangerous feedback loop.[39]

Wildfire smoke isn't just outside—it can get inside your home. There are health precautions you can take if you live in an area impacted by wildfire smoke.

Preparing for Wildfire Smoke

When it's smoky outdoors, try to stay indoors. Remember, smoke from outside can enter your home and affect the indoor air quality as well.

· Close the fresh-air intake if your HVAC system or room air conditioner has one.

· Ask a heating, ventilation, and air-conditioning (HVAC) professional what kind of high-efficiency filters (rated MERV 13 or higher) you can use in your home's HVAC system.

· If you have to shelter in place, choose a room with no fireplace and as few windows and doors as possible, such as a bedroom.

· Plan to use a portable air cleaner in the room.

· Spend as little time outside as you can.

· If you must go outside, wear a mask.

The CDC advises:
· Take it easier during smoky times to reduce how much smoke you inhale. If it looks or smells smoky outside, avoid strenuous activities such as mowing the lawn or going for a run.
· Use the recirculate mode in your car's air conditioner.
· Clean yourself with soap and water as soon as possible if exposed to ashy air.

(Source: EPA)[40]

Wildfire smoke is often visible, but some air pollution threats are not. In cities, another gas called *ozone* is a common threat. Ozone is considered "good" when it's high in the atmosphere. But ground-level ozone occurs when sunlight and heat mix with chemical pollutants from industry or transportation emissions and can irritate the lungs' lining, causing damage. It can reduce lung function and make it harder for you to breathe. Research has linked ozone to asthma attacks.[41] Hazy, hot, and humid weather may call for local "ozone health alert days." Climate change may lead to an increase in ozone.[42]

Air Pollution: Minorities and Poor Suffer More

The burden of air pollution is not evenly shared. Poor people and minorities are among those who often face higher exposure to pollutants and who may experience more significant health impacts.[43] Racial and ethnic differences in asthma frequency, illness, and death are connected with poverty, city air quality, substandard housing, indoor allergens, insufficient patient education, and poor health care.[44] Various studies detail the following inequities:

- Racial minorities and low-income households are disproportionately likely to live near a major road where transportation-related pollution is typically highest.[45]
- Exposure to NO_2 levels generated from the traffic-related population was 37 percent higher in communities of color compared to predominantly white neighborhoods.[46]
- Low socioeconomic status consistently increased the risk of premature death from fine particle pollution among 13.2 million Medicare recipients studied in the most extensive examination of particle pollution-related mortality nationwide.[47]
- Puerto Rican children are twice as likely to have asthma as non-Hispanic whites.[48]
- African American children have the highest prevalence of asthma.[49]
- About 13.4 percent of African American children have asthma, compared to about 7.4 percent of white children with asthma.[50]

- African Americans are three times more likely to be hospitalized for asthma.[51]

Dr. Lakiea Wright is an allergy and immunology specialist at Brigham and Women's Hospital in Boston. She's treated patients with allergies and asthma and has authored research papers on air pollution's inequitable impacts. Dr. Wright shares her expertise and insight in the conversation below.

1. **Can you talk about the racial and income disparities studies that indicate the impact of air pollution?**

   Dr. Lakiea Wright: I think this is interesting. Disparities have been persistent among races, but there are also low-income disparities . . . so if you're a minority group *and* you're low income, it's almost like a double whammy. That may be a symptom of a larger disease. We've been having more of a conversation about structural racism and how historical practices and cultural norms have perpetuated.

   In the Black Women's Health Study, an extensive study looked at perceived racism. It correlated higher experience of perceived racism associated with greater odds of developing asthma as an adult.

2. **How do these factors lead to developing asthma?**

   We know that if you believe you are a victim of racism, that can be associated with stress. Stress has inflammatory changes. We see that there are these biological consequences of perceived racism. This sort of stress can disrupt your immune system. Allergies and asthma are inflammatory diseases, so does it have this sort of synergistic effect? We need more data, but what it suggests is that there may be a connection there.

3. **What are some of the challenges you've seen for low-income families dealing with asthma?**

Access to affordable medications continues to be an area that we study and struggle with. The cost of many inhalers is still on the patents. They are expensive, but there are some generic versions of inhalers. If patients were ever having problems getting their asthma medication, some pharmaceutical companies have programs these patients can access. I found them, as a practicing physician, to be very helpful to be able to refer my patients. They have either no copay or a smaller copay.

I've worked in low-income communities. They didn't realize the rights that they have. You have the right to live in standard housing. When it's substandard, it affects your health. As a physician, I can document your medical condition. I can't make your landlord do something, but I can document it. I've personally had lots of success with that.

The home environment plays with asthma in ways that shouldn't be underestimated. I always like to talk to my patients about their medications and point out what precautions and actions they can take outside of medicine. For example, you can reduce your exposure, and we know that based on clinical studies. Reducing exposure helps improve symptoms and reduce medication use.

4. **What are some resources for low-income families?**

For low-income individuals, they should talk to their health care providers about different resources that may be available. In some instances, it may be possible to visit with a social worker, case manager, or community health worker to understand what resources are available in their community.

There's a healthy homes program with grant funding allocated to different state agencies or nonprofits which can offer home visits to evaluate environmental hazards and provide services that may remediate deficiencies.

More information can be found at https://www.cdc.gov/nceh/publications/factsheets/HealthyHomesInitiative.pdf and https://www.hud.gov/program_offices/healthy_homes/hhi.

5. **Are you seeing disparities in gender for asthma as well—particularly in children?**

Yes, we do see it. It's highly prevalent in African American boys. In general, though, boys may outgrow it more than girls as they age. In some people, though, asthma can go undiagnosed. So we need to get a better sense of how many females out there have asthma. It can be harder to quantify.

6. **What can parents do to mitigate asthma risk for children?**

This is important. Allergies can trigger up to 80 percent of asthma. I think parents often don't realize that for children. It can take a couple of seasons to be entirely noticeable. At around age three or so, you can see children develop symptoms, and their parents say, "Well, I don't know what's going on. They've never had this before."

I would say that you definitely want to talk to your doctor about it, especially if they have allergies and asthma.

Americans spend up to 90 percent of their time indoors. Indoor allergens and irritants play a significant role in triggering asthma attacks. Triggers are things that can cause asthma symptoms, an episode or attack, or make asthma worse.[52]

Asthma in Children

The causes of childhood asthma aren't fully understood. Some factors thought to be involved include inherited tendency to develop allergies, like having parents with asthma, experiencing some types of airway infections at a very young age, and exposure to environmental factors, such as cigarette smoke or other air pollution.

· Asthma is more common in children than adults.

· Asthma is the most common chronic condition among children.

· 3.5 million children suffered from an asthma attack or episode in 2016.

· In 2017, one in twelve children had asthma.

· Currently, there are about 6.2 million children under the age of eighteen with asthma.

· It is the top reason for missed school days. In 2013, about 13.8 million missed school days were reported due to asthma.

(Source: Mayo Clinic and the Asthma
and Allergy Foundation of America)[53]

7. What are your tips to reduce indoor exposure to allergens?

First, let me start by saying if you have allergy symptoms, you want to be able to identify your allergen by talking to your doctor and getting the proper diagnosis.

Identify your triggers: keep a diary of your respiratory symptoms and potential triggers. Review with your doctor, and in some cases, allergy testing may be appropriate for common environmental allergens (which is available in both skin and blood testing).

If you notice that pollen and certain types of weather are triggers, keep aware of outdoor pollen and monitor outdoor air quality. Try to avoid going outside during storms and on days when the air quality is poor. If you are allergic to pollen, try to minimize your time outdoors during the early hours of the morning (five a.m. to nine a.m. are when pollen counts are typically highest).

If you keep your windows open, allergens will come in. A HEPA filter can zap up some of those allergens and even zap out some indoor air pollution. If you buy one, make sure you select it based on your room's size, which many people don't realize.

You also want to create a safe environment in your bedroom. When you are asleep, you're inhaling for eight hours in one spot. If you're allergic to dust mites, washing your bedding regularly and vacuuming, as well as dusting your blinds or curtains, will help.

If you're allergic to pollen or mold, try to avoid going outdoors before thunderstorms. Consider staying indoors and keeping your windows closed. If you must go out before a thunderstorm, consider wearing a mask.

Indoors, other things can serve as irritants, like cooking oil or a gas stove. You want to make sure your stove is vented correctly and well maintained. Replace those HEPA filters based on what your manufacturer recommends. Be aware of cleaning products like air freshener that contains bleach or ammonia. Some of my patients even combine them, not realizing that that's a *toxic* combination. Use more naturally based cleaners without harsh chemicals.

More Indoor Air Quality Improvement Tips for Allergy and Asthma Sufferers

Monitor sources of indoor air pollution, including:

- Fuel-burning heat sources (such as a wood-burning stove)
- Smoke from cooking, candles, fireplaces, or tobacco
- Toxic fumes that are "off-gassing" from new products (new furniture and new carpet)
- Attached garages that store cars, motorcycles, or lawnmowers (can add carbon monoxide to your air)
- Building and paint products (paints, adhesives, solvents)
- Pesticides (such as treatments for cockroaches and fleas)
- Radon (a gas that comes from the ground and enters a home and can rise to dangerous levels)
- Humidity that allows mold to grow
- Cosmetics, perfumes, and hair sprays

Reduce your home's indoor air pollution:

- Remove or reduce any known allergens
- Do not smoke tobacco products in your home
- Prevent mold growth by lowering the humidity in your home by using exhaust fans in kitchens, bathrooms, and laundry rooms (or open a window if necessary)
- If you live in a humid area, consider getting a dehumidifier for your home
- Increase airflow to give your house better ventilation (open windows, doors)

- Store harmful products in a shed that is not attached to your home
- Avoid using scented candles or odor-hiding fragrances
- Install and check carbon monoxide alarms and radon alarms

(Source: Asthma and Allergy Foundation of America)[54]

To Prevent Mold Growth in Your Home

- Keep humidity levels in your home as low as you can—no higher than 50 percent—all day long. An air conditioner or dehumidifier will help you keep the level low. You can buy a meter to check your home's humidity at a home improvement store. Humidity levels change throughout the day, so you will need to check the levels more than once.
- Be sure the air in your home flows freely. Use exhaust fans that vent outside your home in the kitchen and bathroom. Make sure your clothes dryer vents outside your home.
- Fix any leaks in your home's roof, walls, or plumbing so mold does not have moisture to grow.
- Clean up and dry out your home entirely and quickly (within twenty-four to forty-eight hours) after a flood.
- Add mold inhibitors to paints before painting. You can buy mold inhibitors at paint and home improvement stores.
- Clean bathrooms with mold-killing products.
- Remove or replace carpets and upholstery that have been soaked and cannot be dried right away. Think about not using carpet in places like bathrooms or basements that may have a lot of moisture.

(Source: CDC)[55]

6.

SHINING LIGHT
ON WINTER BLUES

The gray, sparsely lit days of winter can be dismally cold. For some, this may mean cozying up to a fire with a good book. But for others, reduced sunlight in January and February can trigger fatigue, despondence, insatiable carbohydrate cravings, and even severe depression. Women are especially at risk, as they are likely to more be sensitive to weather changes and experience depression at twice the rate of men.[1]

Light itself is remarkable. Science suggests it can be used therapeutically to help heal sadness and illuminate the spirit transcendentally through meditation and mindfulness.

Sunlight, or lack of it, doesn't just steer mood. It's also the driving force for our circadian rhythm, the body's twenty-four-hour internal clock. Disruptions—such as when we "fall back" for daylight saving time (initially started as an effort to save energy)—can be detrimental to that natural balance. There are ways, however, to mitigate these potential problems through science-backed hacks that will help you optimize your body's innate timing in any season.

Oh, Those Summer Nights

Renowned psychiatrist and author Dr. Norman Rosenthal grew up in a place filled with sunshine. Johannesburg, South Africa, has a climate comparable to that of San Diego, California—mainly sunny, dry, and mild. When he moved to America as a young man back in 1976 for a psychiatric residency in New York City, Dr. Rosenthal experienced four seasons for the first time. He marveled at the colors of autumn leaves; savored the scent of spring blooms. It was the summertime, though, that Dr. Rosenthal cherished the most. Filled with brightness, the summer days stretched so long that they seemed endless. These seasonal extremes were nothing like his homeland. Located close to the equator, Johannesburg receives approximately equal hours of sunlight both in winter and summer.

During his first summer in New York City, Dr. Rosenthal said he felt boundless energy. But that euphoria didn't last. Soon the air grew brisk, and the long summer nights waned. By late fall and into his first US winter, he noticed his mood changing. He was less cheery and energetic. The cold, gray days and dark, early nights made him feel less productive and not as inclined to be out and about in the city. The slow fade of the summer sun cast a long shadow over Dr. Rosenthal's good mood.

Did Other People Feel This Way?

Dr. Rosenthal began investigating the science behind these seasonal mood swings years later, at his fellowship with the National Institute of Mental Health in Bethesda, Maryland. "When I started doing research anecdotally, I didn't realize how many people in some ways were affected by seasonal depression," Dr. Rosenthal said. "I mean even look at the language: 'a *sunny* disposition, the *sunny* side of the street, *dark* mood versus *bright* mood.' It's in the language, seemingly metaphorical. Yet it was actually *physiological*."[2]

Dr. Rosenthal went on to become an internationally recognized expert on seasonal mood changes. In 1984 he led the research team that first coined the term *seasonal affective disorder (SAD)* as a type of psychiatric depression.[3] This term is common today, but at the time, it was not well received by his peers.

"My thesis was met with a great deal of skepticism in the medical community. Although some people supported this new diagnosis, others were dismissive—contemptuous, even. Some colleagues even thought SAD was a joke,"[4] he recalled to the UK's *Daily Mail*.

During the decades that followed, that perception changed. Seasonal affective disorder is now recognized as a "recurrent major depressive disorder with a seasonal pattern usually beginning in fall and continuing into winter months."[5]

In 2013, Dr. Rosenthal wrote *Winter Blues: Everything You Need to Know to Beat SAD*, currently in its fourth edition. Now a clinical professor of psychiatry at Georgetown Medical School, Dr. Rosenthal explains that seasonal depression is not one size fits all. There is a spectrum. "Nationwide, about 14 percent of people experience it in a mild form, and about 6 percent in a severe way," he said in an interview for this book.

Dr. Kelly Rowan, PhD, of the Department of Psychological Science at the University of Vermont, is conducting a five-year randomized clinical trial, funded by the National Institute of Mental Health, in which 160 community adults with seasonal affective disorder will be treated with either her new SAD-tailored cognitive-behavioral therapy or with standard light therapy and followed over two years. In an online discussion with the American Psychiatric Association, she pointed out a common misconception about SAD.

"I think it surprises a lot of people to hear that really the only thing that makes SAD different from garden variety depression is the seasonal pattern."[6] Rowan explained how the symptoms are similar: "feeling persistently down or sad—losing interest or pleasure and things that a person would normally enjoy like their hobbies or social activities; significant weight changes, which could be a weight gain or a weight loss. Change of appetite, either up or down. Sleep disturbances could take the form of insomnia—difficulty falling or staying asleep. Or the opposite—what we call *hypersomnia*—sleeping too much," Dr. Rowan said in her interview. "Feeling pretty persistently tired and fatigued seems to be the universal symptom."[7]

Aside from the symptoms detailed by Dr. Rowan, other signs and symptoms of SAD may include feeling hopeless and having difficulty concentrating. However, not every person with SAD will experience all of these symptoms.

Cold Nights and Warm Pizza

Some people with SAD report experiencing severe carbohydrate cravings. Research indicates that people with SAD may have reduced brain chemical serotonin activity,[8] which helps regulate mood. Certain foods—including bread, cookies, pasta, white potatoes, bagels, pizza, chocolate, ice cream, and french fries, to name a few—can promote rapid serotonin intake. These effects, though, tend to be extremely short-lived and do not provide your brain with a feasibly sustainable source of serotonin. And this isn't a healthy solution, as it can lead to—you guessed it—weight gain.

In a study cited by the National Institute of Health, researchers say, "serotonin seems to be involved in the abnormal regulation of mood and food intake that underlies diet failures in individuals who suffer from . . . seasonal affective disorder (SAD) . . . characterized by episodic bouts of increased carbohydrate consumption and depressed mood."[9]

What Causes SAD?

Scientists don't fully understand what triggers SAD. Research suggests that sunlight controls the levels of molecules that help maintain normal serotonin levels. And in people with SAD, this regulation does not function properly, resulting in decreased serotonin levels in the winter.[10]

Serotonin is considered one of the four "happy hormones" in the brain, the others being dopamine, oxytocin, and endorphins. All of these can help promote positive feelings and good moods.[11]

Where you live can be a factor in determining if you are at risk for seasonal depression. Unsurprisingly, it's more common in northern latitudes.[12] At Yale University, located in New Haven, Connecticut, researchers write: "SAD has a history of affecting the Yale community. Due to Yale's northern latitude and frequency of gray days, seasonal affective disorder impacts students."[13]

Another study, by scientists from Brigham Young University (BYU) published in the *Journal of Affective Disorders*, took the research one step further. They paired meteorological data with psychological information and found that people who lived in regions with shorter, darker days were more likely to experience more emotional distress.[14]

At BYU, an analysis of past weather variables such as wind chill, rainfall, solar irradiance, wind speed, and temperature data was combined with mental and emotional health stats from BYU's Counseling and Psychological Services Center to look for correlations.

The team presumed that cloudy and rainy weather would correlate with more mental stress, but what they found was that the amount of time between sunrise and sunset was the factor that matters most.[15] "That's one of the surprising pieces of our research," said Mark Beecher, a clinical professor and a licensed psychologist in BYU's Counseling and Psychological Services. "On a rainy day or a more polluted day, people assume that they'd have more distress. But we didn't see that. We looked at solar irradiance or the amount of sunlight that actually hits the ground. We tried to take into account cloudy days, rainy days, pollution . . . but the one really significant thing was the amount of time between sunrise and sunset."[16]

The BYU researchers said their findings indicate that therapists may find a higher demand for counseling during the winter months.[17]

The Hormone of Darkness

Other research suggests that people with SAD produce too much *melatonin*, a primary hormone for maintaining the normal sleep-wake cycle. Overproduction of melatonin can increase sleepiness.[18]

Melatonin is a naturally occurring hormone in the brain, secreted by the *pineal gland*, also deep in the brain. "It's the hormone of darkness, so when the ambient light goes down, there's something called dim light melatonin onset," said Dr. Qanta Ahmed, MD, attending sleep disorder specialist at NYU Langone Health Systems.

"As the light goes down, the brain detects the light levels falling and starts to secrete pulses of melatonin, and the brain says, 'Oh good, it's dark, it must be time to get ready for sleep,' and begins to ready the brain for sleep onset."

Dr. Ahmed says some of her New York–based patients travel to Florida annually in the winter to help regulate their sleep patterns and manage seasonal depression. "The winter here in the Northeast—we're in northern latitudes, so our days are shorter, and we have fewer hours of daylight than in the summer," she said. "Some people find it very difficult to feel alert in the early dark mornings, and they see the sun going down at four or five p.m., and it makes them feel tired and sleepy."

Dr. Ahmed, who often appears nationally on television as a renowned medical expert and commentator, provides more examples of an environment where light is lacking and how it affects human behavior.

"If you visit an elderly relative or friend in a nursing home or an assisted living facility," Dr. Ahmed said, "most elderly people are spending all of the time indoors and not exposed to bright light. So you will go there and visit them, and nine times out of ten, they will be asleep in the day. That is because of the lack of exposure to sunlight. They may also have other sleep disorders, but the lack of exposure to sunlight in institutionalized Americans causes daytime tiredness and daytime sleepiness."

Both serotonin and melatonin help maintain the body's daily rhythm tied to the seasonal night-day cycle. In people with SAD, the changes in serotonin and melatonin levels disrupt the normal daily rhythms. As a result, they can no longer adjust to the seasonal changes in day length, leading to sleep, mood, and behavior changes.[19]

The Sunshine Vitamin

Over a billion people worldwide are deficient in vitamin D,[20] which can be consumed through diet, or absorbed through skin exposure to sunlight. Low levels of this nutrient have been reported as a significant risk factor for many chronic illnesses.[21] A deficit may exacerbate seasonal depression, as studies indicate it's a factor in promoting serotonin activity.[22]

An international research partnership between the University of Georgia in Athens, the University of Pittsburgh, and the Queensland University of Technology in Australia discussed their findings on vitamin D deficiency and seasonal depression in the November 2014 issue of the journal *Medical Hypotheses*.[23]

Alan Stewart of the University of Georgia College of Education, an associate professor in counseling and human development, told Sciencedaily.com, "Vitamin D levels fluctuate in the body seasonally, in direct relation to available sunlight. For example, studies show there is a lag of about eight weeks between the peak in intensity of ultraviolet radiation and the onset of SAD, and this correlates with the time it takes for UV radiation to be processed by the body into vitamin D."[24, 25]

Women Are More Affected Than Men

Of the millions of Americans who suffer from SAD, three out of four are women. Between 60 percent and 90 percent of people diagnosed with SAD are women between fifteen and fifty-five.[26] Why do women bear the brunt of this? There's evidence that ties rising and falling levels of female hormones like estrogen and progesterone with depression.

For those who've experienced *premenstrual syndrome* (PMS), this may sound familiar. The increased risk of depression for females "is associated with fluctuating estrogen levels that occur during reproductive cycle events like periods, post-pregnancy and particularly during *perimenopause*—this transition is a time characterized by drastic fluctuations in estrogen levels and increases in new-onset and recurrent depression."[27]

Research shows symptoms impact women most during their reproductive years. Dr. Norman Rosenthal speculates this has to do with the "cyclical reproductive hormone session that is interacting with the sensitivity of the brain, possibly the hypothalamus or some other part of the brain is light sensitive." He added, "You might even sort of go a little further along the speculative line. There are reasons for seasonal rhythms of reproduction in the animal kingdom. Different species give birth to their young when the young have got a better chance of survival." Dr. Rosenthal points to gestation and reproductive cycles that coincide with weather and climate during different times of the year.

Women were more likely to become pregnant in the summer and thus give birth at a time of year when their babies had a higher chance of survival.[28] Historically, "winter depression symptoms also promoted healthier pregnancies and . . . improved the survival chances of both mothers and babies,"[29] according to research by English psychiatrist Dr. John Eagles. He hypothesized that during the winter, "Enhanced female-male pair-bonding resulted in better health outcomes for mothers and babies." Dr. Eagles surmised that increased energy and elevated moods often occurring in spring and summer also "served to increase the likelihood of procreation at the optimal time of year. In the modern era, it is probable that recurrent winter depression is becoming a reproductive disadvantage."

According to the National Institute for Mental Health, there are other risk factors for developing the most severe type of seasonal depression—SAD. These include people who also have a major depressive disorder or bipolar disorder. SAD also sometimes runs in families. Seasonal affective disorder is more common in people who have relatives with other mental illnesses, such as major depression or schizophrenia.[30]

Light Is *Zeitgeber*

Just a few generations ago, most of the world's population was involved in agriculture and was outdoors for much of the day. Before the invention of artificial light, natural sunlight guided the day. Our ancestors were exposed to high levels of bright light even in winter.[31]

"If we look at the beginning of the twentieth century, there wasn't this enormous amount of electricity that we have everywhere. People actually woke up with the crows at sunrise and would go to bed at sunset," Dr. Ahmed said. That's not the case today, as most people spend 90 percent of their time indoors, surrounded by artificial light.

"We actually lose our entrainment to sunlight, our knowledge of sunrise and sunset," said Dr. Ahmed. "In the course of a century, Americans are losing almost ninety minutes of total sleep time just because we have so much artificial light."

Dr. Ahmed explained that during the short days of winter, many people work in offices without windows. "We don't actually have a relationship to the time of day outside. That can produce its own sleep disorders."

Light is measured in *lux*, a unit of illumination in the International System of Units (SI), from the French Système International d'Unités. Lux, which is Latin for light, is the measure of radiant light from a standard candle that falls on one square meter of the surface area one meter from the source.[32]

Even on a cloudy day, the light outside can be greater than 1,000 lux, a level never normally achieved indoors.[33] One study carried out at around latitude 45° N found that daily exposure to light greater than 1,000 lux averaged about only about thirty minutes in winter and ninety minutes in summer[34] among people who worked full-time.[35]

"As biological organisms," Dr. Ahmed explains, "we are programmed to

How Is SAD Diagnosed?

If you think you may be suffering from SAD, talk to your health care provider or a mental health specialist about your concerns. They may have you fill out specific questionnaires to determine if your symptoms meet the criteria for SAD.

To be diagnosed with SAD, a person must meet the following criteria:

· They must have symptoms of major depression or the more specific symptoms listed above.

· The depressive episodes must occur during specific seasons (i.e., only during the winter months or the summer months) for at least two consecutive years. However, not all people with SAD experience symptoms every year.

· The episodes must be much more frequent than other depressive episodes that the person may have had at different times of the year during their lifetime.

If you or someone you know is in immediate distress or is thinking about hurting themselves, call the **National Suicide Prevention Lifeline** toll-free at 1-800-273-TALK (8255) or the toll-free TTY number at 1-800-799-4TTY (4889). You also can text the **Crisis Text Line** (HELLO to 741741) or go to the **National Suicide Prevention Lifeline** website, https://suicidepreventionlifeline.org.

(Source: National Institute of Mental Health)[36]

wake up and sleep almost entirely in relation to light. Light is the most power-ful source or what we call *zeitgeber*, a time giver that tells us that it's time to wake up or time to fall asleep."

Light is also linked with our emotions, according to Dr. Ahmed. "It affects mood tremendously. If there's bright light—a normal sunny day, for example, is measured at a light strength of about 10,000 lux, equivalent to 10,000 candles. Most artificially illuminated buildings are operating somewhere at about 400 to 1,000 lux—much, much dimmer, so we're not mimicking the exposure to sunlight that we can."

Even a one-hour change in available light makes a difference. That's why, for many, the shift of one hour during daylight saving time can offset their biological clock or circadian rhythm.

The Energy of Daylight

The concept of daylight saving time (DST) is often credited to Benjamin Franklin, who famously espoused the virtues of "early to bed, early to rise." But according to History.com, "he only proposed a change in sleep schedules—not the time itself."[37] The idea of actually moving the clock is traced back to Europeans in the early twentieth century and was implemented in the US as a wartime measure in 1918.

Daylight saving time was initially advocated as a means to *conserve energy*. But the data didn't always point to substantial improvements. A US Department of Transportation study in the 1970s concluded that total electricity savings associated with daylight saving time amounted to about 1 percent in the spring and fall months. As air-conditioning has become more widespread, more recent studies have found that cost savings on lighting are more than offset by significant cooling expenses.

Also according to History.com, "University of California, Santa Barbara, economists calculated that Indiana's 2006 move to statewide daylight saving time resulted in a 1 percent rise in residential electricity usage. This increase appeared to be caused by additional demand for air-conditioning on summer evenings and morning heating in early spring and late fall. A US Department of Energy report found a decrease in energy use per day during extended daylight saving time, but only by 0.5 percent. Some also argued that increased recreational activity during DST resulted in greater gasoline consumption."[38]

In a 2007 study, researchers concluded that "contrary to the policy's intent—DST *increases* residential electricity demand." Their findings pointed toward higher uses for both heating and cooling, which outweighed the reduced potential benefit of lighting power.[39]

Hawaii and Arizona—except the state's Navajo Nation—do not observe daylight saving time. The US territories of American Samoa, Guam, Puerto Rico, the Virgin Islands, and the Northern Mariana Islands also remain on standard time year-round. Some Amish communities also choose not to participate in daylight saving time.

Daylight saving time has come under scrutiny in recent years, with various legislators moving to abolish it in their states. The European Union voted to eliminate it in 2019. In 2020, New York City mom Lisa Singer started a petition to persuade New York State governor Andrew Cuomo to permanently get rid of daylight saving time in New York State.

"My eight-year-old daughter's sleep was disrupted every year—I noticed how exhausted she felt and how it negatively affected her," Singer said.

The daylight saving time debate continues in many states. If you live in a location where each year you "fall back" and "spring forward," there are steps you can take to make the adjustment easier.

Regulating Rhythm

Dr. Ahmed says exposure to natural light in the morning is key to regulating circadian rhythm. If you're up and out of the house before dawn, look for that light when you can, even if you're stuck inside.

"My advice is that anyone working in a building or being in the home, try to get to a window, look up at the blue sky, get into some natural light. If you are struggling to feel alert and awake in the morning, make sure that you can sit next to a window, go outside, don't wear your sunglasses. Skip them for an hour or two so that you get connected to the external light," she advises.

Dr. Ahmed also mentioned that even brief exposure to sunlight in the afternoon can assist with energy slumps between lunch and dinner. "If you're

Health Tips to Prepare for Daylight Saving Time

- Start preparing a few days to a week early. Go to bed fifteen to thirty minutes earlier or later than your usual bedtime.
- Stick to your schedule. Be consistent with eating, social, and exercise times during the time transitions.
- Avoid coffee and alcohol four to six hours before bedtime. Alcohol also prohibits you from getting quality sleep, so avoid it late at night.
- Don't take long naps lasting longer than twenty minutes.
- Expose yourself to the bright light (preferably natural) in the morning.

(Source: Cleveland Clinic)[40]

in an office building, try to go out on your lunch break to get some fresh air so that you're not in a building with recirculated air and also have five to ten minutes in daylight, and you'll notice an enormous difference. It will help you get through the afternoon at the office."

Since lack of sleep can exacerbate the symptoms of seasonal depression, it's important not to hinder the biological process when it's time to wind down. That's why you want to avoid artificial light before bed. Scrolling emails and social media on your phone or laptop before sleep exposes your eyes to blue light, and that will keep you up when you don't want to be.

"When you're exposed to any device, even if you're looking at it in the dark, whether it's a television or a mobile phone, that is immediately transmitting light through the *retinohypothalamic track*," Dr. Ahmed said. This track is the pathway for light levels detected through the eyes to be transmitted to the brain's hypothalamus, the region that controls the body's twenty-four-hour internal clock, also known as the circadian rhythms.[41]

As tempting as it may be, reaching for your phone in bed to read texts sends the wrong signal to the brain. "The suprachiasmatic nucleus thinks, 'oh it must be the day,'" Dr. Ahmed explains. "This can promote insomnia. If you're looking at the devices before you go to bed, you're watching the eleven o'clock news, you're watching your late-night comedy, your brain thinks it's still daytime, and it delays the arrival of sleep. Simple rules, and just to put it in a nutshell, the bedroom is for sleep and sex only. No other activity belongs there."

Lighting the Path to Improved Mood

Exposure to light can also be therapeutic for those who suffer from seasonal depression. It's one of the most common treatments for SAD. (Others include antidepressant medications, talk therapy, or some combination of these.) While symptoms will generally improve on their own with the change of season, they can improve more quickly with treatment.[42]

Light therapy involves sitting in front of a light therapy box that emits a very bright light (and filters out harmful ultraviolet rays). The light box is a fluorescent box from which the patient sits about ten feet away, and it may deliver 10,000 lux at a time. Doctors treating SAD have advised patients to sit in front of the box for twenty to forty minutes or an hour each morning, and have found it improves mood.

Dr. Rosenthal was one of the earliest experts to utilize these devices in treatment. "I described seasonal affective disorder in the mid-1980s, and we did that early work with light therapy. And now it's been validated across the world that more light helps people, especially in the morning," said Dr. Rosenthal.

You don't need a prescription to get a light box, but experts advise they are best used under the guidance of a doctor. Some people may begin light therapy in the early fall to prevent symptoms when winter descends.

Dr. Rosenthal said that the amount of time patients spend in front of the light box can vary depending on the time of year they use it. For example, less time may be needed in November when days are just beginning to get shorter, and more time in January when nighttime darkness is more prevalent.

The doctor also advises not to skimp on size when looking for an artificial light box. "The smaller ones are very seductive. They're cheaper. They're easier to accommodate in a smaller space, but they could very well be less effective," he warns.

Many who use these boxes see some improvements as early as within one or two weeks.[43] Studies indicate when used correctly and consistently under medical supervision through the winter months, between 50 percent and 80 percent of light-therapy users have complete remissions of symptoms.[44] It's advised to consult your physician before you begin to use a light box.

Lifestyle Tips to Manage Seasonal Depression[45]

- **Eat healthy:** Get creative and look for hearty, low-calorie recipes that are easy to prepare.
- **Avoid social isolation:** Spend time with your friends and family, snuggle with pets. Friends and family can be good to talk to about how the season is affecting you.
- **Stay active:** Volunteer, join a local club, go for a walk, be proactive about planning activities in advance during the winter to keep active and engaged with others.

The Mind-Body Connection

During the winter, people with seasonal affective disorder have a reduced ability to handle stress, which can push them deeper into depression, according to Dr. Rosenthal. "I recommend that you do whatever you can to minimize stress. This means thinking about winter ahead of time. For example, do not undertake projects with a spring deadline attached to them, because you know ahead of time that that will put you under stress during the winter."[46]

Dr. Rosenfeld is a huge advocate of leaning into mindfulness and meditation. "It has been enormously helpful in managing my own SAD symptoms since regularly practicing Transcendental Meditation [TM], and many others have found this practice equally helpful." This interest inspired Dr. Rosenthal to write *Transcendence: Healing and Transformation Through Transcendental Medita-tion* in 2011, a *New York Times* bestseller. "TM had done so much good both for my patients and myself, had such strong research backing, and was so pleasant and easy to do that I felt the urge to share these experiences as seen through the eyes of a doctor, scientist, and TM practitioner," Dr. Rosenthal said. He still practices this form of meditation daily.[47]

The TM technique involves using a silent sound called a mantra, and it's practiced for fifteen to twenty minutes twice per day.[48] (To find a certified TM teacher in your area, visit the website TM.org.)

How to Do Transcendental Meditation Step by Step[49]

1. Look for a quiet environment to perform your meditation.
2. Find a comfortable position for you to sit for at least thirty minutes.
3. Relax, close your eyes, and breathe deeply.
4. Repeat your chosen mantra or Sanskrit sound. Transcen-dental meditation teachers may provide students with a sound, like "Om" (pronounced "Ohm"; it means "it is," "will be," or "to become," and is often referred to as "the universal mantra").[50] Another chant to consider is "Lu-men de Lumine" (Latin for "of the light, for the light").[51]
5. If you feel yourself getting distracted, repeat your mantra or sound to get back into the present and into your medi-tation.
6. After twenty minutes or so have passed, ease back into your surroundings by rolling your shoulders or moving your toes and fingers.
7. When ready, open your eyes slowly. You're done with your session.[52]

(Source: learnrelaxationtechniques.com)

7.

WEATHERING AUTOIMMUNE FLARE-UPS

I t's hard to imagine our own body attacking itself. But that's at the root of more than eighty different autoimmune diseases that cause the body to assault its own healthy organs, tissues, and cells. Autoimmunity is now one of the most common disease categories, ahead of cancer and heart disease. And while rates of the latter are falling, autoimmune diseases are being diagnosed with such frequency that some medical experts are calling it an epidemic.[1]

For many, it's a long journey to confirm the cause of their painful and debilitating symptoms. Statistics show patients face an average of five years and five different doctors to receive an official diagnosis of an autoimmune disorder, according to the American Autoimmune Related Diseases Association (AARDA)—even though some 50 million Americans suffer from one.[2]

Women are worse off on this front as well; they're three times more likely to develop an autoimmune disease—like rheumatoid arthritis, for example—than men. To add to that burden, research shows that those living with an autoimmune disease are often more likely to suffer from more than one illness at the same time.

"There's still a lot of mystery associated with autoimmune disease," says Kathleen Gilbert, an immunologist and retired professor at the University of Arkansas for Medical Sciences. Cures have yet to be discovered.[3]

Climate change has been associated with an increase in autoimmune disease.[4] Nitrous oxide, one of the leading greenhouse gases contributing to global warming, has been noted by researchers to be a correlating factor.[5] Pollution containing particulate matter, when inhaled into the lungs, triggers the body's immune system into reacting protectively to kill abnormal cells. This inflammatory response is part of the reason air pollution aggravates allergies and asthma.

Researchers believe the lung itself may be an initiation site in autoimmune diseases.[6] As reported in cantechletter.com, a new Canadian study suggests a link between air pollution and autoimmune diseases like rheumatoid arthritis and lupus.[7]

Symptoms for many with autoimmune disease don't remain constant; they can intermittently range from inactive to excruciating. An *autoimmune flare-up* is a temporary period of worsening and intensification of symptoms due to an added stressor.

Sometimes that stressor can be weather-related. Meteorological phenomena or environmental-related conditions triggering flare-ups are routinely relayed by patients to their doctors.

Anecdotally, you may have heard someone say when bad weather is coming, they can "feel it in their bones." When a storm approaches and barometric pressure drops, aches and pains, especially for those with arthritis, are common. Less expectedly, though, a bright sunny morning may signify a debilitating day is ahead for those with autoimmune diseases like lupus. Bright UV rays could trigger rashes and fatigue.

Lupus disproportionately affects women of color, who are at greater risk of developing the disease. For those with psoriasis, the winter's cold and dry climate can instigate itchy skin.

There are precautions you can take each season to ease symptoms of autoimmune diseases triggered by certain weather conditions and the environment.

Bombing Out with Pain

Watching a weather report, you may hear the meteorologist describe a "low-pressure system forming off the coast and ramping up in intensity as it spins counter-clockwise and edges northward." During the winter, storms like these, swirling with cold air and wrapped in fierce winds, may be referred to as nor'easters—named for their harsh northeastern winds. Inside their low-pressure centers, rising air lifts rapidly, and atmospheric pressure plummets. Sometimes the storm's pressure falls so fast over twenty-four hours that the storm is said to be *bombing out*. The mere mention of this meteorological term fills weather geeks with excitement. However, for the millions who have joint inflammation from arthritis, a bombing-out nor'easter may not be as thrilling, as these intense winter storms can trigger an explosion of pain!

Does Science Support the Weather Connection?

Mumbai-based rheumatologist Dr. S. M. Akerkar says, "Many of my patients keep telling me that weather affects their arthritis on a regular basis. Most of them complain that cold or damp weather and rains make them feel worse than sunnier, warmer, drier weather. Whether it's true has been a hotly debated topic."[8]

According to Dr. Akerkar, there are theories as to why drops in barometric pressure may cause joint pain. When atmospheric pressure (the weight of the surrounding air) drops, this allows tissues in the body to expand to fill up space. "The inflamed *synovium*—soft tissue that lines the joints and tendons' spaces—is hypothesized to swell and cause more pain. Another explanation is that damp and cold weather causes the muscles to contract as the body shivers to maintain body temperature. This traction on the joints may lead to pain."[9]

Note that the term "arthritis" is an informal way of referring to joint pain or joint disease. There are more than one hundred types of arthritis and related conditions. People of all ages, sexes, and races can and do have arthritis, and it is the leading cause of disability in America.[10]

Other factors associated with cold weather that may cause a physiological response like joint pain involve the constriction of blood vessels. For example, in freezing temperatures, vessels narrow and restrict blood flow, which can cause discomfort.

Cloudy with a Chance of Pain

How do weather conditions synch up with pain propensity? That's what a fifteen-month Versus Arthritis study of over 2,500 participants across the UK set out to discover. Researchers said that relative humidity appeared to affect patients with *osteoarthritis* more than those with other types of weather-related conditions, though they concluded that "most of those studied are sensitive to the weather."[11]

Osteoarthritis is the most common type of arthritis. It's not classified as autoimmune in cause, as it occurs when the *cartilage*—the slick, cushioning surface on the ends of bones—wears away and bone rubs against bone, causing pain, swelling, and stiffness. Over time, joints can lose strength, and pain may become chronic. Risk factors include excess weight, family history, age, and previous injury.

For arthritis sufferers, even subtle weather changes can lead to varied perceptions of pain. Another study examined patients across four US cities: San Diego, Nashville, Boston, and Worcester, Massachusetts. The researchers anticipated receiving higher reports of joint pain for those who lived in New England due to the greater variety of weather there, but that was not the case. Their findings suggested that it was those living in the consistent sunshine and mild temperatures of Southern California who were most sensitive to minute weather fluctuations.[12]

In general, while there's no universal scientific consensus, medical experts say weather extremes of any kind will place additional stress upon the body, which is usually not helpful for those suffering from autoimmune diseases. Storms, freezing temperatures, and sudden drops in barometric pressure can increase the incidence and severity of autoimmune conditions in a non-specific way simply by adding to the heightened physiological demands of the body during such periods.[13]

This stress can leave an autoimmune sufferer more susceptible to flares, which *might* be better controlled in more temperate weather, though this varies for each individual.[14]

CreakyJoints.org is a digital community for millions of arthritis patients and caregivers worldwide that offers information and support. Dr. Anca Askanase, a rheumatologist and director of rheumatology clinical trials at Columbia University Medical Center, says less movement in the winter season may be an additional factor that contributes to greater discomfort that season. "One of the reasons that your pain may seem worse when it's cold is because people are less inclined to move or go outside when it's chilly."[15] She recommends doing some light exercise, like yoga, Pilates, tai chi, or qigong, or taking a short walk to help with joint pain.

Tips to Cope with Arthritis in the Winter

• **Stay Active:** Exercise is crucial for people living with arthritis. Despite the misconception that it will aggravate symptoms, physical activity actually helps ease pain, increase strength and flexibility, and boost energy. Adults with arthritis who have regular physical function and no other severe health conditions are advised to do at least 150 minutes of moderate-intensity aerobic activity a week, as well as two weekly sessions of strength training.

   On days when the very thought of exercise seems unbearable, gentle range-of-motion exercises can be helpful.

• **Maintain Balance:** Choose footwear that provides stability and traction. Stay on cleared sidewalks and paths if possible. If you need more stability, consider a cane, walker, or even trekking poles to help keep you steady on your feet.

   If arthritis has caused you to have balance issues, try mall walking, doing workout videos online, or consider a gym membership.

• **Keep Warm:** Heat is like a spring thaw for your stiff joints. It boosts blood flow to help flush out pain-producing chemicals and stimulates receptors in your skin that improve your pain tolerance. Warmth also relaxes muscles to decrease spasms and reduce stiffness.

• **Try Heat Therapy:** A warm shower or a soak in the tub (dress warmly afterward to prolong the benefit), a heating pad (opt for one that delivers moist heat, which penetrates more deeply than dry heat), electric blankets, single-use hand and feet warmers that you can slip into your gloves, pockets, or shoes, or even wrapping your hands around a hot cup of coffee or tea and warming up the car before hitting the road can be helpful in your quest to quell joint pain.

- **Wear Compression Gloves:** You might want to try thermal compression gloves. They vary in style and technology, but some will emit or help trap heat to provide a warming sensation and provide compression to reduce swelling. Most styles are fingerless, so they're not suitable for outdoors unless you slip another pair of gloves or mittens on over them. If you don't have a strong preference, mittens tend to be warmer than gloves because your fingers generate more heat when they're not separated from each other by fabric, as they are with gloves.

- **Avoid "D" Deficiency:** It can be challenging to get enough vitamin D in the winter. It's produced when your bare skin is exposed to sunlight. Even though your body can store vitamin D levels throughout the seasons, research shows people are often deficient in the winter. That can spell bad news for inflammatory arthritis patients. Low vitamin D levels are also associated with pain sensitivity. Vitamin D is one of a few supplements for arthritis doctors often recommend.

    Talk to your doctor about testing your vitamin D levels to be sure you're not deficient (some arthritis drugs can interfere with its absorption). If your vitamin D level is low, you may be advised to take a supplement.

(Source: Creakyjoints.org)[16]

Psoriasis

Cold and dry weather can also trigger outbreaks of *psoriasis*, a chronic auto-immune disease that can generate scaly or red, itchy areas on the skin. The cause of psoriasis is not known, but researchers believe both genetics and environmental factors play a role. Because of its appearance, psoriasis is often believed to be contagious, but that's a misconception; it isn't. For the approximately 4.5 million people who suffer from it, though, it can be debilitating.

Psoriasis is also stressful because it's unpredictable. It can lie dormant and then flare up at almost any moment. Itchy scales can erupt on various parts of the body, and flare-ups range from mild to severe.[17]

In winter, psoriasis can be at its worst. Cold temperatures and dry air can strip away the thin layer of oil that traps moisture on the skin's surface, leading to flare-ups of flaming, itchy patches.

What Does Psoriasis Do to the Skin?

Under normal conditions, the top layer of our skin sheds and grows back over twenty-eight to thirty days.[18] In patients with psoriasis, however, this natural process is sped up considerably; cells are regenerated in as little as two to three days. When this happens, the old cells cannot keep up. They remain even while new cells are multiplying. This causes the cells to stick together and form lesions or patches that are called *plaque*. They are typically dry, scaly, inflamed, and sometimes itchy. The more dehydrated your skin, doctors say, the worse those patches will look and feel.[19]

People with *psoriasis arthritis*—a form of arthritis that affects some people who have psoriasis—also experience joint pain and plaque in flares. About 30 percent of people with psoriasis also have psoriatic arthritis.[20] This inflammation of the joints can be painful. And like arthritis, psoriatic arthritis flares, for some patients, can also be triggered by cold weather. Winter can be twice as woeful with this double whammy. Some describe the chillier temperatures as making it feel as though "the fluid in their joints seems thicker."[21]

Stress is one of the most common psoriasis triggers. At the same time, a psoriasis flare itself can cause stress.[22] This may seem like an endless loop. However, relaxation techniques and stress management may help. Here are some tips for managing skin flares from psoriasis in the winter.

Dermatologist Dr. Navya Handa's Tips to Manage Psoriasis During Winter[23]

(Check with your doctor before starting any home treatment.)

- **Apply moisturizer to keep your skin moist.** Washing hands and bathing remove oils, so apply at least twice a day to keep your skin well hydrated. Products containing urea or salicylic acid can help soften and remove dry-skin buildup in affected areas.

- **Opt for a lukewarm bath** instead of a hot shower. Long showers in hot water can strip the moisture from your skin.

- **Use a humidifier.** Indoor humidity levels during the winter can drop to 20 percent. Adding moisture to the air will help to raise that level to over 40 percent, which is beneficial.

- **Wear soft clothing layers.** Fabrics like wool and synthetic fibers can irritate your skin. Choose soft cotton clothes as the first clothing layer instead.

- **Bundle up to stay warm** but be mindful of overheating. Sweating, especially on the face or scalp, can irritate your skin and cause a psoriasis flare-up.

- **Ensure proper nutrition and hydration.** Omega-3 fatty acids can help reduce skin dryness. Drink enough water, as it helps the skin to retain natural moisture.

- **Reduce stress levels.** Stress is a potential trigger for psoriasis; practice yoga, meditation, and other relaxation techniques.

- **Consider swimming for exercise**, particularly in saltwater, as it may offer some relief.

- **Maintain optimal vitamin D levels** to help prevent psoriasis flare-ups.

UV Light Treatment for Psoriasis

Sunlight can not only boost your mood but there are also physical benefits for psoriasis sufferers. "If you ask most dermatologists, they'll tell you that a great many of their patients flare in the winter," says Dr. Abby Van Voorhees, a dermatologist at Eastern Virginia Medical School, in an interview with psoriasis. org. "Our theory is that a diminished amount of ultraviolet light is behind the flares. This is why artificial UV light is hugely helpful."[24]

Doctors harness the power of natural sunlight as a treatment for psoriasis. This is implemented through *phototherapy*, or light therapy containing ultraviolet B (UVB) rays. When penetrating the skin, UVB rays can slow the growth of affected skin cells. Initially the treatment is done under medical supervision, after which patients use a light unit at home. When doing phototherapy at home, experts advise people to follow a health-care provider's instruction and continue with regular checkups.[25]

The excimer laser, approved by the Food and Drug Administration (FDA), is also used for treating chronic, localized psoriasis plaque. It emits a high-intensity beam of UVB. This device can target areas of the skin affected by mild to moderate psoriasis. Research indicates it is a particularly effective treatment for scalp psoriasis, which affects 80 percent of patients. However, there is not yet enough long-term data to show how long improvements last after a course of laser therapy. All phototherapy treatments, including purchasing equipment for home use, require a prescription from your health-care provider.[26]

UV Light, Poor Air Quality, and Lupus

While UV therapy can be beneficial to people living with psoriasis, for many with *lupus*, another autoimmune disease, the sun's rays can lead to painful flares. Lupus, technically known as systemic lupus erythematosus (SLE), is an auto-immune disease in which the body's immune system mistakenly attacks healthy tissue in many parts of the body.

Sunshine emits light and heat. The heat itself doesn't cause this adverse reaction for those with lupus; rather, the sun's invisible UV rays do. Health experts say this increased sensitivity to sunlight is known as *photosensitivity*. As many as 75 percent of lupus patients are photosensitive.[27]

UV exposure can trigger flare-ups of this autoimmune disease. One of the most common lupus flares patients display is the *butterfly* or *malar rash*, which stretches across the nose and the cheeks in a butterfly shape. Other skin problems include disk-shaped lesions and scaly red circles, plus symptoms like fatigue and joint pain. In more severe cases, this could also lead to fever and organ inflammation.[28]

"UV light damages skin cells and causes their death, thus exposing the inner parts of cells to the immune system—which, in susceptible individuals, can lead to an autoimmune reaction," said Dr. George Stojan, assistant professor of medicine at the Johns Hopkins Lupus Center.

Sources of UV light can be found beyond the obvious places, like a sunny day at the beach. Indoor fluorescent lighting, copy machines, and sunlight passing through a car window can all trigger a flare. Certain medications can also make you more sensitive to light.

When forty-nine-year-old lupus patient Ellen Schnakenberg ventures outside on a bright day, her skin breaks out in a sunburn-like rash from UV exposure. Then she gets so ill that she has to retreat to her bed. Schnakenberg told Lupus.org: "I'll have a flare that will last for a couple of days to several weeks."[29]

Lupus impacts women, minorities, and low-income communities disproportionately.

Stats: Women, Minorities, and Lupus

- Women develop lupus much more often than men: nine of every ten people with lupus are women.

- Lupus is three times more common in African American women than in Caucasian women.

- As many as 1 in 250 African American women will develop lupus. Lupus is more common, occurs at a younger age, and is more severe in African Americans.

- Hispanic, Asian, and Native American populations are also more likely to develop lupus.

(Source: Lupus.org)[30]

It's unknown why lupus is more common in African Americans. Some scientists think that it is related to genes.[31] A 2010 study by a team of researchers in Cambridge, UK, found that a specific gene variant is strongly linked to an increased risk of lupus. The variant also makes a person more resistant to malaria, which occurs in African and Asian regions of the world. This finding indicated this ancestral gene would be helpful and selected for in areas of the world where malaria is prevalent. "In those places, the downside of increased lupus risk would be far outweighed by the added protection against malaria."[32] Differences in the quality of health care available for lupus among racial and ethnic minorities and those living in poverty may reflect more limited health-care access.[33]

Hormones, age, and environmental factors additionally play a role in who may develop lupus. Reproductive-age women are disproportionately impacted, according to a Canadian study that found an increased risk of lupus in women using oral contraceptives.[34] Other research has indicated that exposure to trace elements—especially heavy metals—prevalent in the environment can be triggers for disease onset. These may include uranium, lead, cadmium, mercury, nickel, and gold. The latter "have potentiate delayed hypersensitivity reactions in patients with connective tissue disease. . . . The resultant enhanced inflammation has been considered a pathogenic factor in exacerbations of autoimmune and inflammatory diseases including Lupus."[35]

The Weather/Climate Connection to Lupus

Baltimore-based rheumatologist Dr. Stojan is one of the leading experts and researchers in fighting the battle against lupus. He observed a correlation between weather conditions and reported autoimmune symptoms through years of working with many patients. To further investigate, he matched EPA data on temperature, wind, humidity, barometric pressure, and air quality with reported symptoms from over 1,900 patients in the ten days before their doctor's visit.

"Science over the past hundred years has mostly been concerned with how the immune system works as an isolated system—without influences from the exterior," Dr. Stojan said. "I think that this research and data does show that we as organisms are affected by where we live, along with environmental changes."

His findings reflected these weather/symptom connections:

- Rising temperatures: joint swelling, organ tissue inflammation, rashes, and declines in red and white blood cells and platelets
- High humidity: joint swelling and inflammation
- Poor air quality: inflammation, new rashes, blood abnormalities, and joint pain[36]

"I think that this research does bring the possibility that the increased exposure to pollutants certainly raises the risk of some organ-specific flares for people with autoimmune disease," Dr. Stojan said.

Exposure to outdoor air pollution depends on each individual's location, the magnitude of the components, and their everyday activity patterns. In cities, for example, time spent closer to roads with heavy traffic raises risk.

Pollutants are known to cause oxidation damage to the lungs.[37] The body's effort to repair this triggers inflammation, which can aggravate autoimmune diseases, causing flares. "The interplay of genetic background and environmental factors could potentially explain the high incidence and increased severity of lupus in some ethnic minorities in the United States, like African Americans and Hispanics,"[38] says Dr. Stojan.

Dr. Stojan said that poor air quality is a public health issue that officials need to consider when making policies. "We might need to make changes in the acceptable levels of exposure in the air of nitrogen dioxide, sulfur dioxide, fine particulate matter—because I don't think that the current recommendations take into account the fact that these minor shifts can affect the health of patients with autoimmune disease," he said. "If there is the political will and knowledge to make the changes to mitigate that risk . . . I think that would make a big difference in the long run."

Tips to Reduce Lupus Photosensitivity

Many people with lupus are sensitive to light. People of *all* skin colors are at risk for skin damage and many other harmful effects of ultraviolet radiation from the sun, and for some, from indoor fluorescent or halogen lighting. For those with lupus, UV exposure can trigger photosensitivity and the activation and worsening of systemic symptoms—including joint pain and kidney disease. Here are some helpful things to know about UV light to help limit your exposure.

1. UV radiation is classified by wavelength—the distance between the peaks in a series of light waves. UVA is the longest wavelength. It penetrates deep into the middle layer of the skin and can cause long-term damage. UVA is a danger every day, regardless of weather, location, or altitude (even passing through window glass). UVB varies in intensity depending on the weather and season. It's

stronger in summer and more intense at midday, at high altitudes, and near the Equator.

2. The hours between ten a.m. and two p.m. are when the sun's UVB rays are most intense. Whenever possible, plan outdoor activities in the early morning and the late afternoon (and try always to wear sun-protective clothing and apply broad-spectrum sunscreen no matter what time of day it is).

3. UV radiation is more intense when reflected by water, sand, and snow, in locations closer to the sun (such as mountains), and in countries near or at the Equator.

4. Cloud cover and shade do not provide complete protection from UVA.

5. Medications may increase the body's sensitivity to the sun—talk to your pharmacist or doctor about your increased risk if you're taking antibiotics, anti-inflammatory or blood pressure medications, or methotrexate.

6. If limiting your UV exposure prevents your body from making enough vitamin D, eating foods with vitamin D and taking vitamin supplements can help you meet your daily requirement. Talk to your doctor about your vitamin D levels.

7. Fluorescent and halogen light bulbs emit UVA radiation that can cause flares of lupus in people who are especially photosensitive. Plastic blocks both UVB and UVA, and plastic light shields are available for many types of indoor bulbs—you may want to request them as a necessary accommodation at your workplace.

8. Tanning is actually the unhealthy response the body has to repair skin cell damage. Artificial tanning beds are especially dangerous for photosensitive people, as the light wavelength used is primarily UVA, which penetrates deeply into the skin.

(Source: Lupus.org)[39]

8.

HEAT, SLEEP, AND MEMORY

Heat, it turns out, is not just a sticky haze that makes us binge on electricity to keep the air conditioner running. It interferes with daily routines, makes us less successful at cognitive tasks, and generally lowers productivity. Most people aren't aware of this, blaming themselves for their poor performance on hot days. Heat's impact on our mind is even more magnified when our body cannot cool down at night. In this uncomfortable external environment, our sleep is stolen and our mental performance suffers the next day.

Extreme heat is projected to increase due to climate change. Climate Central estimates that the average American will experience between four and eight times as many days above 95 degrees by the year 2030. What may surprise you is that overnight low temperatures are going up even faster than daytime highs. Research points to these diurnal anomalies stealing an increasing amount of precious sleep in the decades to come.

Air-conditioning isn't always guaranteed, as we've seen recently with power outages and rolling blackouts. There are things you can do, though, to help keep cool and sleep better at night to improve your chances of maintaining daytime mental sharpness.

Sweltering for Science

As uncomfortable as it sounds, living, sleeping, and taking daily morning cognitive tests for weeks in an oven-hot apartment was embarked upon willingly by college students, all in the name of science. In the first study of its kind, healthy young people (previously only elderly adults were studied) would be evaluated for the effect of extreme heat on their mental acumen.

Dr. Jose Cedeno Laurent, PhD, of the Harvard School of Public Health, wanted to determine if more significant portions of the general population were susceptible to mental impairment due to extreme heat.

On the Harvard University campus in the summer of 2016, the living quarters of forty-four students were divvied up distinctively. One half lived in modern six-story buildings with central air-conditioning. The other half were housed in low-rise concrete buildings constructed over half a century ago, with no AC or even fans to cool them. All of this occurred during the hottest days of the year.

"We spend 90 percent of our time indoors, yet most homes and buildings are not constructed to withstand extreme heat," said Dr. Cedeno Laurent. Older housing was designed to keep the cooler air outside so residents could stay warm. This is particularly true for New England, which is better known for harsh winters and snowstorms. Keeping cold air out was the priority. But during stretches of hot summer weather, inside these sturdy cold-busting dwellings, something else was happening. Even after it cooled down relatively outside, inside it remained just as warm. Dr. Cedeno Laurent calls this phenomenon an *indoor heat wave*.

The students experienced this effect firsthand. At the time of the Harvard study, for five days, 90-plus degree temperatures held Cambridge, Massachusetts, and the surrounding areas in a stranglehold. Then a sharp cold front sliced through the outside steamy air, and outdoor temperatures dropped down to the 70s. The community at large breathed a huge sigh of relief. For the college kids stuck inside the hot dorms, though, the change was barely noticeable. The indoor heat wave kept them sweltering.

At night, instead of peaceful sleep, they tossed and turned between sweaty sheets. Each morning, as the students blearily opened their eyes, they groggily reached for their smartphones. Their first task of the day was to take cognitive tests delivered to their mobile devices.

This routine went on for weeks. When the study was finally over, Dr. Cedeno Laurent anticipated he'd see performance differences between students living in the hot and cold residences, but he was shocked by the extent of them. "To our surprise, we found a very significant effect of detrimental cognitive function," he said. The students who didn't have air-conditioning in their dorms did worse, by at least 13 percent, on both the cognitive and math tests than the students who lived in air-conditioned dorms.[1]

The Harvard study proved extreme heat could slow down your thinking and make your short-term recall less accurate, no matter how old you are.

Why? It's all rooted in brain science.

Heat on Your Mind

Weighing about three pounds, the brain is the control center for just about every physiological process in our bodies. Much like a thermostat in our home, the brain's hypothalamus regulates our core temperature. Its priority is to keep the body from overheating. When external heat rises, our hypothalamus directs more blood to the skin to evacuate heat. That's priority number one, and it's often performed at the expense of other internal systems.

"By shifting more of that cognitive load to dealing with hot temperatures, you take away some of the brain's cognitive role," Dr. Cedeno Laurent said.

When the brain is preoccupied with keeping our bodies cool in the unrelenting heat, its capacity to remember a word or solve a math problem is reduced.

During the study, extreme heat hindered the students' cognitive function not just at the time they took their morning tests but also each preceding night by stealing their slumber. Sleep is one of the prime drivers of keen mental performance; achieving restorative rest is intrinsically tied to the temperature of your surrounding environment.

Degrees of Sleep

As night falls, our body temperature naturally drops—dipping to its lowest level in the hours before dawn. We must cool for sleep to commence and for it to be sustained through the night.

"Our sleep typically will be disrupted if we're not in that nice, ideal temperature zone," explains Dr. Markus H. Schmidt, MD, PhD, a senior sleep researcher at the University of Bern, Switzerland. "We are very good at creating our own little microclimate, which we do in our bed with our covers," he said. "If it's too hot, we may put a foot out, or an arm out, or uncover, etc., so we can radiate the heat and keep the temperature low. Or, we just add a lot of covers if we're cold. We tend to thermoregulate very nicely behaviorally."

The optimal ambient temperature for sleep varies from person to person, but experts generally recommend a range between 60 and 67 degrees Fahrenheit to be most comfortable and effective. That number may need to be adjusted to an even cooler range for some women, though, according to NYU Winthrop and Langone Hospital Attending Sleep Disorder Specialist Dr. Qanta Ahmed, MD.

"If you ask any perimenopausal woman who is struggling with menopause, menopause involves a vasomotor response," she said. These are symptoms that occur due to the constriction or dilation of blood vessels. "It comes from an intense release of heat, so-called hot flashes, associated with sweats. It wakes the patient up from sleep," she explained.

"It causes trouble falling and staying asleep because of that disruption in the control of body temperature due to menopause or perimenopause. This can account for insomnia in those individuals, and they particularly need to keep the room cooler to facilitate sleep onset," Dr. Ahmed said.

In these instances, the London-born, New York–based doctor pointed out that finding the mutually agreed upon, ideal sleep temperature for men and women who share a bed can be particularly challenging. "If I talk to my perimenopausal patients, some of them want the bedroom at 55 degrees. Obviously, their husbands are freezing and wearing wooly hats if they're co-sleeping with their partner," she said.

When the bedroom temperature is too warm, not only is it tough to fall asleep, it's also more challenging to delve into brain-benefiting restorative sleep.

Sleep on It

At night we toggle between two quite different stages of sleep. At the start of snoozing, we begin with non-rapid eye movement sleep (NREM), where our breathing is slow and regular, and our limbs or eyes move minimally. During these NREM stages, the brain also sorts through various experiences from the previous day, filtering out important ones and eliminating other information. These selected memories will become more concrete as deep NREM sleep begins. This process will continue during rapid eye movement sleep (REM),[2] which occurs approximately ninety minutes later, when the memory is thought to be more fully consolidated. This lasts about ten minutes each hour, progressively getting longer until it lasts an hour before waking up.

REM is a paradoxical state where our breathing becomes fast and irregular, our limbs twitch, and our eyes move rapidly from side to side.[3] In this stage, our brain is highly active; images, sounds, and other sensations are relayed from our five senses and integrated into dreams. The brain transmits these cues to the cerebral cortex, a thin layer of the *cerebrum*—the most prominent and uppermost portion of the brain.

Recent research has concluded that during REM, specific neural activity occurs that is required to consolidate memories.[4] At this stage, as busy as our brains are, our bodies become paralyzed and can no longer naturally thermoregulate. Meaning you can't sweat during REM.

"No one understands why. We sweat in non-dream sleep, we get to dream sleep, we stop sweating," explains Dr. Schmidt. "This loss of thermoregulation in REM sleep is one of the most peculiar aspects of sleep."

Releasing heat through sweat is crucial to our survival. It allows us to maintain a healthy inner body temperature. It also requires a lot of energy, meaning the process itself is taxing on the body. When we're asleep in a bedroom that is too warm and the ninety-minute cycle reaches the REM point, the brain makes an immediate choice: "REM sleep is the first to go," Dr. Schmidt said, meaning the beneficial REM process disengages as the brain overrides it, allocating its full energy to regulating our internal body temperature. That subconscious choice has its consequences. Depriving the mind of that stage of sleep, when our eyes are darting from side to side and the brain is consolidating and storing memories, may be one reason why a cognitive function can decline the next day without it.

"We've learned that sleep before learning helps prepare your brain for the initial formation of memories," says Dr. Matthew Walker, a sleep scientist at the University of California, Berkeley. "And then, sleep after learning is essential to help save and cement that new information into the architecture of the brain, meaning that you're less likely to forget it."[5]

Harvard researchers say that memories begin to accumulate during the day, moment by moment, while you're awake. "When we first form memories, they're very raw and fragile," says sleep expert Dr. Robert Stickgold of Harvard Medical School. "Sleep seems to be a privileged time when the brain goes back through recent memories and decides both what to keep and what not to keep. During a night of sleep, some memories are strengthened."[6] Research has shown that memories of certain procedures, like playing a melody on a piano, can improve while you sleep.

Scientists don't know exactly everything that occurs during REM, but it's clearly essential for us. Skimping on it can make us feel less alert the next day. Not sleeping or getting enough sleep can lower your learning abilities by as much as 40 percent.[7]

This is what happened to the students living in the hotter Harvard dormitories. When it came time for REM during their sleep cycles, their brains prioritized internal cooling efforts and, therefore, wouldn't allow it.

Dr. Cedeno Laurent's results put it plainly: no air-conditioning leads to less sleep, which in turn leads to lower cognitive test scores. "Our study's participants all had wearable devices to measure total sleep time throughout the experiment. We did see shortenings of that time. During the heat wave, there was an increase in the awake/restless periods for those with no air-conditioning."

The Harvard study proved that extreme heat could mess with your mind and memory.

Losing Sleep over Climate Change

When the summer sun sets, we anticipate a nice, cool reprieve from the heat of the day, but that's proving to be something we can't always count on. The trend of increasingly warmer nights is evident in climate data. Over the past century, looking back on the average low temps over the past century, the nighttime temperatures have seen the most significant acceleration over the past fifty years. Since 1970, average summer nights in the US have warmed by 1.8 degrees.[8]

These numbers are enough to keep you up at night—figuratively and literally.

In 2017, the most extensive study conducted by a team of researchers, led by University of California San Diego's Dr. Nick Obradovich, PhD, examined the relationship between sleep and ambient temperature. Their goal was to answer the question "Will climate change—through increases in nighttime heat—disrupt sleep in the future?"[9]

In order to solve their query, the team gathered past data from 765,000 sleep survey respondents recorded over the course of a decade. The researchers investigated the relationship between climatic anomalies and insufficient sleep accounts noted by participants, then predicted how their findings might coincide with projected temperature rises due to climate change.

The team discovered substantial correlations between atypically high nighttime temperatures and insufficient sleep. Using the data from survey respondents along with prior climate information from NASA, the researchers determined that a one-degree Celsius deviation from average monthly nighttime temperatures creates an additional three nights of disrupted sleep for every hundred people.

To put that in perspective: one month of nightly temperatures averaging one degree Celsius above average would equate to roughly 110 million extra nights of insufficient sleep annually.[10]

These findings showed that just a tiny nudge upward of the nighttime thermometer impacts sleep deprivation, aligning with the study Dr. Cedeno Laurent conducted at Harvard. "We noted that the increase of one degree Celsius in overnight temperatures resulted in a three-minute decrease in our study participants' total sleep time," Dr. Cedeno Laurent said. "Their worst scores on the tests matched up with the hottest nights and the lack of quality sleep that was achieved the night before."

The University of California group looked to determine what this research might mean for future sleepless nights on a warming planet. Using climate projections for 2050 and 2099 by NASA Earth Exchange, the study paints a bleak picture of the future if the relationship between warmer nights and disrupted sleep persists. Warmer temperatures could cause six additional nights of insufficient sleep per a hundred individuals by 2050 and approximately fourteen extra nights per one hundred by 2099. Areas of the western and northern United States—where nighttime temperatures are projected to increase most—may experience the most considerable future changes in sleep.[11]

Cooling While Warming

The solution to keeping the bedroom cool for optimum sleep may seem simple during hot weather: air-conditioning. And it's a popular one, especially in the US. In fact, up to 85 percent of America's population lives in air-conditioned homes. But the century-old technology also has a major flaw: it is actually helping to create *more* extreme heat.

Air-conditioning relies on refrigerants and chemicals that can readily absorb heat from the surrounding environment. As they soak up and then cast out hot air, these chemicals move from the inside to outside repeatedly. The problem is that the most commonly used refrigerants in residential air-conditioning are among the most potent greenhouse gases on Earth—a thousand times as intense as carbon dioxide. This creates a feedback loop that is cooling your home while helping to raise global temperatures at the same time.

There's also the added energy consumption with running an air conditioner. The heavy use of air conditioners stresses the power grid, which often leads to outages. The electric grid's carbon intensity, the amount of carbon dioxide released per unit of electricity generation, also goes up. In the US alone, approximately the same amount of electricity is used just for air-conditioning as the total amount used for all needs by 1.1 billion people in Africa.[12]

While most homes have air-conditioning, the distribution varies by city and state. For example, only about 35 percent of residents in states like California, Washington, and Oregon own air conditioners. Longtime inhabitants of those cities would likely tell you that the heat wasn't so bad in years past, and they generally didn't need them. But that may be changing. Cities like Seattle, Portland, and San Diego have seen about five more days of 85 degrees or higher temperatures in the past few decades. These typically cool places have all reported their hottest summers on record in recent years.

At the University of California, San Diego, the lack of air-conditioning inspired Nick Obradovich's large sleep/temperature study. As a graduate student there, he found that he was sleep-deprived and in a perpetual bad mood after several days of extreme heat and living and sleeping with no air-conditioning. As Obradovich ambled around campus, sweating and exhausted, he noticed everyone around him looked as tired and bedraggled as he felt.[13]

His scientific findings of future warmer sleepless nights do not come as good news to the current state of widespread, national sleep deprivation. The CDC reports that currently, in America, 70 percent of adults get insufficient sleep at least one night a month, and 11 percent report inadequate sleep every night.[14]

Natural Cooling

Longer and more frequent heat waves lead to REM-robbing nights. Plus, air-conditioning's negative feedback loop, and the unreliability of an increasingly stressed power grid, warrants the question: What are some hacks to improve the odds of successful sleep in case AC is not available?

Health experts emphasize that before you even focus on naturally cooling your bedroom, the process of achieving good quality sleep starts when you wake up in the morning. This is because the body's circadian rhythm, or internal twenty-four-hour clock, is triggered by light. "As biological organisms, we are programmed to wake up and sleep almost entirely in relation to light," Dr. Ahmed said.

This process begins when light enters our bodies through our eyes, traveling through an internal pathway straight to our brain's center and into the hypothalamus. Inside this tiny power-house of neurons, light triggers the master circadian clock of the mind, generating an internal representation of solar time conveyed to every cell in our body. Your circadian rhythm directs our behavior and physiology's daily cycles, which set the tempo of our lives.[15]

Dr. Ahmed recommends exposure to at least an hour of direct sunlight in the morning. That way, your brain can register the light to start the day. This will boost morning energy and lead to more restorative sleep that night.

But connecting with the morning light is not always easy. Many people commute to work early in the morning before the sun rises and go straight inside to insulated offices. And at higher latitudes, we experience shorter days and longer nights.

That fluorescent glow isn't helping. Indoor lighting, even at its brightest, can't compensate for the real thing, especially in windowless offices. As explained earlier by Dr. Ahmed, "A normal sunny day is measured at a light strength of about 10,000 lux, equivalent to 10,000 candles. Most artificially illuminated buildings are operating somewhere at about 400 to 1,000 lux—much, much dimmer, so we're not mimicking the exposure to sunlight that we can.

"So, my advice is," Dr. Ahmed continued, "that anyone working in a building or being in the home should try to get to a window and look up at the blue sky, especially in the morning, get into some natural light." Even spending fifteen minutes outside in daylight or next to a bright window can make a difference in aligning circadian rhythm for better sleep.

Keeping Cool with a Warm Shower

The same circadian rhythm ignited by light in the morning to wake us up also signals our body temperature to rise to about 2 to 3 degrees Fahrenheit higher in the late afternoon. Yet we must internally cool down in order to fall asleep.

It may sound counterintuitive, but had the students in the Harvard study taken a warm shower before bed, they might have slept somewhat better and, in turn, possibly scored higher on his cognitive tests.

Research shows that one way to help your body cool down before sleeping in an unairconditioned bedroom is to take a warm bath or shower. That's right, *warm*—not cold.

It sounds counterintuitive when it's sweltering outside, particularly if you don't have air-conditioning, as it's more instinctive to opt for a cold bath or shower before bed or even soak your feet in ice! But that's not a good idea, according to Dr. Schmidt.

"If you take a super-cold bath, it will be nice, maybe for a few moments while you're there, but now you're triggering *vasoconstriction*. That means that the vessels to your hands and feet will be circulating less blood, which is the opposite of what you want. You need to *vasodilate*. You need to get the heat out of your core and to the surface. By having a cold bath, you're going to come out of that tub, and all these vessels are vasoconstrictive," he explained. "This would reverse what you just were trying to accomplish when you get out of the bath."

How warm is lukewarm? There's a big difference between tepid and scalding. Researchers at the University of Texas found that taking a shower or bath with a water temperature of between 104 and 109 degrees Fahrenheit improved overall sleep quality. When scheduled one to two hours before bedtime, it can also help you fall asleep an average of ten minutes faster.[16]

The University of Texas study's lead author, Shahab Haghayegh, a PhD candidate in the Department of Biomedical Engineering, was so impressed with his findings that he has put the practice into use himself before attempting sleep in a warm bedroom. "Since the data proves that a warm shower or bath before bed does make a huge difference in your overall sleep quality, I take a warm shower every night before bed now," he wrote in an article following the experience for the Thrive Global community.[17]

Bedroom Chill and Air Circulation

We are sentient beings, but our furniture is not. That much is clear. Chairs, tables, rugs, and beds all absorb heat and, depending on their shape and color, our bedroom can either retain heat or expel it. Darker colors and rougher texture tend to trap heat. Light-colored fabrics and surfaces are cooler.

It's a good idea to keep the shades down, and curtains are drawn during the day to block the sun's heat. For bedding, use breathable fabrics for sheets and linens, like cotton (satin absorbs more heat); this may also help keep you comfortable.

While you may have heard that freezing your sheets or pajamas is a good idea so they will be pre-chilled before bed, that's actually not the best choice. After a while, the bed warms, and that will leave you with damp sheets and bedclothes—leading to more discomfort.

Waiting to Exhale

Ever let out a deep relaxing sigh the moment you get into bed? In an interview with NPR, James Nestor, author of the book *Breath: The New Science of a Lost Art*, says, "The exhale is a parasympathetic response. Right now, you can put your hand over your heart. If you take a very slow inhale in, you're going to feel your heart speed up. As you exhale, you should be feeling your heart slow down. Exhaling relaxes the body."[18]

When we blow out air, it not only feels good but also allows us to release carbon dioxide. Each small puff of breath we expel may seem like a slight amount, but in a well-sealed home, especially in a small bedroom, it can accumulate enough to influence the CO_2 levels. Too much of it, and you may end up with disturbed sleep and a headache the following day. That's what MIT graduate student Joel Jean found when he tested this idea in his own bedroom.

"I've been monitoring the carbon dioxide levels in my bedroom (10.5 ft x 14 ft x 8.5 ft) in Cambridge, MA, continuously over the last two years, using a scale with a built-in carbon dioxide sensor," he wrote in a Medium blog.[19]

"To test the effect of ventilation on indoor CO_2 levels, I alternated sleeping with my bedroom door open and closed for several days. With the door closed—no ventilation—average CO_2 levels increased by over 1,000 ppm [parts per million] during the night." Jean was surprised by the dramatic results.

For the next part of his home CO_2 level test, he cracked the bedroom door open at night, allowing moderate ventilation. This resulted in reduced levels—"roughly 500 ppm during the night."

Improved ventilation can also be achieved by opening a window. Closed windows and doors can reduce the effective ventilation rate, raising CO_2 levels to 2,500 ppm or more, according to additional research published in the journal *Indoor Air*.[20] Before you open your windows at bedtime, remember to check any advisories for poor outdoor air quality, and address safety measures.

Other home heat-reducing hacks include potentially temporarily sleeping in an alternate room and eliminating hidden heat sources. If it's possible to snooze on the lowest level of your home on super-hot days, do it.

The last thing you want to do is make your bedroom warmer. Yet you may not realize there are hidden heat-producing culprits all around you: your computer, laptop, printer, video game console, and television. These electronic devices emit warmth even while they're charging. Unplug them or move them to another part of the house.

Not too surprisingly, in the hour before people try to sleep, technology use is high. Reports indicate that 67 percent of polled Americans use cell phones, 60 percent are on computers or laptops, and 18 percent play video games just before bed.[21] And this electronic activity doesn't stop once we fall asleep. How many times a night do you wake up and check your cell phone, respond to text messages, or just scroll social media?

"If I had to pick any one problem that's affecting Americans more than anything, more than the temperature, more than the carbon dioxide, more than the humidity levels, more even than the sleep disorder of your bed partner, it is the use of artificial devices that is driving so much insomnia today," Dr. Ahmed said.

Blue-light exposure before bed is an additional challenge you don't want to take on, particularly when your goal is to fall asleep in a hot room and then wake up with a cool, sharp mind.

9.

FOOD CHOICES
AND CARBON FOOTPRINTS

Steaming hot coffee first thing in the morning, followed by crisp yet creamy avocado toast. For lunch, fresh lobster and succulent corn on the cob. Sounds delicious! What's for dinner?

So many tasty foods to choose from; it's often hard to pick. Each eating opportunity offers a multitude of options. Restaurant menus list delectable entrees, appetizers, and desserts. Supermarket shelves are fully stocked with name-brand staples, international items, and newly released gourmet treats. At home, it's easy to find yourself lingering in front of an open refrigerator for several moments as you contemplate what you want for a snack.

There's more at stake here, though, than satisfying individual cravings. What we eat, where it comes from, how it's processed, transported, how food waste products are disposed of—all have serious impacts on our planet. The consequences of climate change may limit the seemingly countless choices of things we love to eat.

Take that previously described breakfast of coffee and avocado toast or lunch with lobster and corn, and now add environmental stressors. Extreme heat deteriorates coffee beans, extended drought damages avocados, and windstorms can wipe out thousands of acres of corn. Warming oceans can not only be destructive to lobster fisheries but also force crustaceans to migrate northward in search of a more hospitable habitat.

Different types of plants and animals require varying amounts of precious natural resources (like water) to thrive. Carbon emissions not only pollute the air but can also potentially reshape the nutritional structure of certain harvests. Even staple crops like rice may lose dietary value in vitamins and minerals due to rising levels of CO_2 in the atmosphere. There's already high demand for these harvests. Almost half of our plant-derived calories come from just three sources: wheat, corn, and rice.[1]

For low-income communities in developing countries, climate change is projected to lead to dangerous food insecurity levels.[2]

Innovative solutions in food sustainability are emerging as high tech intersects with traditional farming. In agriculture, these efforts include regenerative soil on cattle pastures, strategic water use, and better food waste management. Farmer's markets are sprouting up in new locations, offering consumers greater access to abundant organic bounties. At home, learning more fruitful and effective ways to shop and store food can reduce what ends up uneaten and thrown away.

So if you consider yourself a "foodie," take comfort in knowing that all is not lost: you can still savor what you love while simultaneously helping the environment. Bon appetit!

Plants and Animals

Land, water, and minerals are all fundamental to how food is grown. These natural resources are under increasing strain.[3] Astonishingly, a mere twelve plants and five animals make up 75 percent of the entire world's food supply.[4] This high demand challenges our food system's sustainability and the long-term health of people and the planet.[5]

Energy is also required to produce, process, package, distribute, and consume food, and to manage its waste products. This is true for foods derived from both plants and animals. All of these actions impact the Earth's biodiversity.[6] This includes all the plants, animals, and microorganisms that keep soils fertile, pollinate plants, purify water and air, keep fish and trees healthy, and fight crop and livestock pests and diseases.[7]

Plant physiologist Dr. Lewis Ziska, formerly with the USDA's Agricultural Resource Service and now an associate professor of environmental health sciences at the Columbia University Irving Medical Center, says that different agricultural products place varying strains on natural resources. "Just to give you an example, to grow a potato from seed to harvest it might take as much as thirty or forty gallons of water. To grow a pound of beef from inception to slaughter will take up to two thousand to three thousand gallons of water," he says.

Water is already a precious commodity, particularly in agriculture. In most world regions, over 70 percent of freshwater is used for farming and food processing. The water demand is forecast to grow along with the rising global population. As reported in veganfoodandiving.com, "By 2050, feeding a planet of 9 billion people will require an estimated 50 percent increase in agricultural production and a 15 percent increase in water withdrawals."[8]

If animals require more water than plants, should we eat less meat? In 2019, the U.N.'s Intergovernmental Panel on Climate Change (IPCC) released a report suggesting eating less meat as part of overall efforts to reduce climate change.[9] Recommendations by the researchers included altering dietary choices to include mainly plant-based foods. "Plant-based foods, such as coarse grains, legumes, fruits and vegetables, and animal-sourced food produced sustainably in low greenhouse gas emission systems, present major opportunities for adaptation to and limiting climate change,"[10] said Debra Roberts, one of the report's authors.

Hans-Otto Pörtner, ecologist and co-director of the IPCC working group on impacts, adaptation, and vulnerability, said: "We don't want to tell people what to eat, but it would be really useful . . . if people in many rich countries consumed less meat."[11]

Interest in and curiosity about how to eat less meat and more plants are reflected in social media. According to Google trends, searches for the term "plant-based" more than doubled over the past five years.

But becoming a full-fledged *vegan*, meaning eliminating all dairy and meat products from your diet, may be further than some are willing to go. A huge survey of public opinion, the People's Climate Vote, distributed various questions to over 1.2 million respondents across 50 countries and released their findings in early 2021. The results indicated that while most people (64 percent) said that climate change was an emergency, altering the way they eat was not a remedy they favored.[12]

The survey asked about eighteen potential policy changes to address climate change in areas including energy, economy, transportation, and food choices. Respondents' answers did not reflect what the polling team anticipated.

Researchers found that just 30 percent of people surveyed were proponents of pushing plant-based diets to prioritize future climate policy. "The promotion of plant-based diets struggled to get even the majority for advocating in any of the countries surveyed."[13] The research team concluded that people believe what they choose to eat is more personal than a political decision.

Another report, titled "Climate Change and the American Diet" and released in February 2021 by the Earth Day Network and the Yale Program on Climate Change Communication, pointed to a lack of knowledge on the subject. This report found that roughly half (51 percent) of Americans surveyed said they would eat more plant-based foods if they had more information about the environmental impact of their food choices. However, the majority of them—about 70 percent—rarely or never talk about this issue with friends or family.[14]

Currently, only 2 percent of the US population is considered vegetarian or vegan. Compared with the number of vegans and vegetarians today, there are more than five times as many former vegans and vegetarians.[15]

Cattle and Carbon

While some advocate for completely eliminating meat production to lower carbon emissions, additional experts would opt for more sustainable methods for cattle farming or focusing on reducing greenhouse gases from other sources of pollution.

"There are all these other things that contribute to climate change, but for some reason, it's the agricultural aspect, in particular beef, that has become the scapegoat," said New York–based board-certified family physician Dr. Gabrielle Lyon, DO. Her global practice focuses on muscle-centric medicine, and she's an expert on dietary protein. "If you think that they're consuming too many resources, you should not own domestic animals; you shouldn't be flying. You shouldn't be using electricity. It would be very hypocritical to then also use leather."

Dr. Lyon points to what she says are misleading public impressions regarding the number of carbon emissions that American cattle actually produce. "According to the Environmental Protection Agency, greenhouse gas from beef cattle represents a much smaller percentage of emissions in the US than the media would lead us to believe," she said.

Dr. Frank Mitloehner, a professor and air quality specialist in cooperative extension in the Department of Animal Science at UC Davis, shares these sentiments. He says livestock does have an impact on the climate (cows that emit methane as they digest food), but "in reality, it doesn't hold a candle to the damage being done by cars, trucks, planes, and industry, or by our insatiable need for electricity. Animal agriculture is a drop in the greenhouse-gas bucket."[16]

In the US, greenhouse gas emissions from the agriculture economic sector accounted for 9.9 percent[17] (this breaks down to about 5 percent from fruits and vegetables, roughly 2 percent from beef cattle,[18] and 2 percent dairy[19]) of the total US greenhouse gas emissions in 2018.[20]

The EPA estimates the primary source of US pollution is transportation. They report that 28 percent of greenhouse gas emissions are by the transportation sector—this includes cars, trucks, commercial aircraft, and railroads.[21]

The *New York Times* reports that in the United States cattle do not comprise the largest source of greenhouse gases, which include carbon dioxide, methane, and others. "Their total contribution is dwarfed by the burning of fossil fuels for electricity, transportation, and industry. But livestock is among the largest sources of methane, which can have 80 times the heat-trapping power of carbon dioxide, although it persists for less time."[22]

According to Stanford University scientists, as reported in nature.com, global methane emissions are up by 10 percent over the past two decades, "resulting in record-high atmospheric concentrations of the powerful greenhouse gas."[23]

On a worldwide scale, livestock accounts for a higher percentage of human-induced total greenhouse gas emissions than in the US, at about 14.5 percent. Roughly two-thirds are attributed to cattle.[24]

Reducing the Carbon Hoofprint

Putting the cows out to pasture. It's a well-known idiom that may have new meaning for finding new ways to reduce cattle emissions that are harmful to the environment. The concept of *regenerative agriculture* in farming draws from decades of research designed to mimic nature.[25]

Technological, genetic, and management changes that have taken place in US agriculture have made livestock production more efficient and less greenhouse-gas-intensive, according to Mitloehner. Oklahoma State University researchers agree that restorative agricultural production practices can decrease atmospheric carbon and reverse some of the effects of climate change.

These enhancements to a farm's entire ecosystem include placing a premium on soil health, strategic water use, and using organic fertilizer.[26] There is minimal to no soil tilling in this practice, crops are strategically rotated, and plant diversity is prioritized to create rich, nutrient-dense soil leading to more productive yields.

Improving the state of soil is key. It's estimated that about one-third of the world's topsoil for agriculture farming is already acutely degraded, making farms less fertile.[27] The United Nations projects this to worsen to total degradation within the next half century under current practices.[28]

Not beneficial to soil health is overgrazing—when too many animals on one pasture feed for too long in one spot.[29] When cows are crowded in the same confined area for an extended period, they chew surrounding plants and grass down to the bare soil. In order to avoid this, farmers are making an effort to managing animals on pastures more efficiently.

Carbon-Neutral Cows

"The newest reports are that grass-fed cattle can actually be neutral to greenhouse gases," said Dr. Lyon. "The gases that the cattle produce are absorbed by the grasses (pastures) they live on. The existence of cattle improves the grasslands' health and gives them the potential to work as *carbon sinks*."

A carbon sink is any reservoir, natural or otherwise, that accumulates and stores carbon-containing chemical compounds for an indefinite period. Number one is the world's oceans. Any guess on where the second-most carbon repositories are found?

If you guessed the Earth's soils, you're right. The world's soils contain an estimated 2,500 billion tons of carbon. The problem is that farmed soils, through modern agricultural practices like plowing and commercialized farming, have resulted in the loss of 50 to 70 percent of the carbon they once contained.

This is where the climate-friendly concept of *soil sequestering*—pulling existing carbon from the atmosphere back into the soil—comes in. It's a practice that some industry observers, as reported in agriculture.com and ew.content .allrecipes.com, say is returning to "a time when lush prairies covered much of the central United States. The grasses then provided food for vast herds of bison and other herbivores that, in turn, fertilized the soil. This symbiotic relationship promoted a vigorous regrowth of vegetation, improved nutrient content, and allowed the ground to hold more moisture."[30]

Regenerative agriculture is gaining popularity. Yale researchers report that soil sequestration has the potential to remove 250 million metric tons or more of carbon dioxide per year in the United States alone.[31] These sustainability efforts garnered investment by major US food companies like Stonyfield and General Mills. Both are running regenerative farming pilots in 2021.[32]

Agricultural regeneration does face challenges, though. One of them is the cost of transition, which remains a commonly cited obstacle among farmers and organizations serving farmers.[33]

Australian cattle farmers remain optimistic based on reported results. Since 2005, the beef industry has reduced its emissions by around 60 percent through incorporating soil sequestering, increasing biodiversity, and changing what cows eat.[34] Did you know when cows eat seaweed and algae, they belch less methane? As reported by ABC News, Meat & Livestock Australia, which regulates meat and livestock management standards, believes a zero carbon footprint (considered by some to be the holy grail for the red-meat industry) in Australia is possible by 2030.[35]

Favorite Foods

The impacts of climate change on different foods and the ingredients needed to make them vary. Whether your favorite meal comes from the land or the sea, the ramifications are already being felt worldwide. Here's a closer look.

Rice

We use it to roll sushi, make a sweet dessert pudding, or cook up a rich risotto. Rice is the staple food for over 3.5 billion people worldwide, who depend on rice for more than 20 percent of their daily calories.[36]

In many developing countries, though, rice is not just a side dish or dessert—it's a primary food source. Rising CO_2 levels are anticipated to be detrimental to the crop's nutrition. The science behind this is rooted in a process you likely studied in a school: *photosynthesis.*

While carbon dioxide can boost photosynthesis and growth in plants, too much of it can also alter their internal chemistry. Studies suggest that the additional carbon dioxide would cause plants to produce more carbohydrates or sugar internally and less vital nutrients like protein, iron, and zinc. Under current emission scenarios, rice could lose the potency of these essential nutrients by almost 20 percent by 2050. These deficiencies could threaten the health of 600 million people globally who rely on rice for most of their daily nutrients.[37]

Corn

Pop it. Slather it with butter. Use it to make sweet-savory bread. Corn is an American staple.

Dr. Eugene Takle, a professor of climate science at Iowa State University, explained in an interview with CNN the various ways weather can help or hurt corn production. Moderate rain can be a boost, but "heavy rains can be detrimental. Those rains pull nutrients out of the earth and cause soil destruction, making it more difficult for corn and other crops to thrive," said Dr. Takle. He also pointed out an increase in the number of warmer, wetter nights in the corn belt that can make it more difficult and expensive to dry corn before it's sold.[38]

Extreme weather events also pose risks to corn crops. One recent example: fourteen hours of straight-line winds that pummeled cornfields along a destructive path of almost a thousand miles! Derechos (that's the name for this type of weather phenomenon, derived from the Spanish word meaning *direct* or *straight ahead*) are not uncommon to Iowa, the leading corn-producing state in the US. But the derecho of August 2020 will not be forgotten.

That's when a powerful derecho traveled from southeast South Dakota to Ohio, a path of 770 miles in over half a day, producing widespread 100-plus mph winds. The storm tore through Iowa as well as Illinois, Minnesota, Indiana, and Ohio. The damage to millions of acres of corn and soybean crops was widespread.[39]

After touring the storm's destructive wake of damage in the days that followed, Iowa secretary of agriculture Mike Naig told Agriculture.com, "Millions of acres of corn around the state were impacted by last week's storm. The severity of the damage varies by field, but some acres are a total loss, and it will not be feasible for farmers to harvest them."[40]

Did climate change play a role in causing the derecho? Scientists say it's difficult to be 100 percent certain. However, there are those who believe the impacts of climate change will increase the intensity of these potentially crop-flattening storms in some areas.[41] And it just takes one decimating derecho to destroy a harvest.

Coffee

Drinking a cup of coffee each morning is often done on autopilot, and it's a popular habit: coffee lovers now consume more than 2.25 billion cups a day! Our changing climate is projected to have negative impacts on coffee production, threatening the livelihoods of millions of farmers globally and affecting the cost and availability of this popular beverage for consumers.[42]

When you stop at your favorite cafe and order a latte or an espresso, keep in mind that there's a very intricate process that makes your morning brew hot and ready right when you need it.

High-end, specialty types of coffees are at greater risk of becoming less available. These highly coveted roasts are often made from plants found at higher altitudes, where there is already a limited amount of arable land. Rising temperatures are projected to "force farmers even farther up the mountainside to seek out the cooler, more desirable coffee-growing conditions found at higher elevations."[43] That's because the higher up the mountain, the firmer the coffee bean will be. Hard beans (as opposed to softer ones) contain stronger concentrations of sugar, which produce more desired and nuanced flavors.

Cooler mountain temperatures also slow the growth cycle of the coffee tree, which prolongs bean development. This longer maturation process imbues the coffee bean with more complex sugars, yielding deeper and more compelling textures. Better drainage at high elevations also reduces the amount of water in the fruit, resulting in a further concentration of flavors.[44]

Nicaraguan coffee, for example, is grown in the most remote of regions: on top of rolling hills and mountains in isolated forests.[45] The highest-grown coffees in Costa Rica might come from farms that are 4,500 feet above sea level, while Ethiopia has farms that sit at 6,000 feet.[46] The constant quest to find this ideal environment results in less available coffee-farming land and can also lead to deforestation.

Natural resources like water are crucial for coffee—and not just to fill your French press each morning. It takes an estimated thirty-seven gallons of water in coffee lands (where beans are grown and harvested) to produce just one cup of joe.[47] Soil to properly cultivate the coffee plant can't be too wet or too dry.[48] Excessive heat can reduce growth and make the coffee plant more susceptible to pests and diseases.[49]

According to research from Conservation International, coffee production will need to increase by as much as 14 million tons of coffee annually to keep pace with demand.[50]

But will there be enough viable land to meet the demand? Current projections by the Intergovernmental Panel on Climate Change estimate reductions in areas suitable for coffee cultivation by 2050.[51]

Sustainable Coffee Drinking Tips

- If you're a K-Pods fan, look for ones that are eco-friendly and made with biodegradable materials.
- Try a reusable coffee cup. This promotes less paper and Styrofoam in landfills.
- Choose fair-trade coffee beans.[52] Fair Trade USA certifies that the farmers were treated and compensated ethically.[53]
- Look for biodegradable coffee filters. Brown, unbleached paper ones are compostable.
- Compost coffee grounds. They are rich in nitrogen and can be helpful to gardening.

Lobster, Salmon, and Other Seafood

Grilled salmon or a lobster roll? These seafood staples are popular, but warming ocean temperatures and ocean acidification may soon put them at risk.

NOAA scientists released the first multispecies assessment of just how vulnerable US marine fish and invertebrate species (think shrimp, crabs, and oysters) are to the effects of climate change. The study examined 82 species that occur off the Northeastern US, where ocean warming is happening rapidly.

The scientists found that the sea creatures most resilient to climate change reside near the ocean's surface, like herring and mackerel. However, fish that migrate from fresh to saltwater—like sturgeon and salmon—are more vulnerable, as are ocean bottom dwellers like mussels, scallops, and clams.[54]

Warmer water temperatures have triggered the northern migration of crustaceans and other creatures, like sea bass, over the past few decades, according to NOAA scientists. In the US, North Atlantic fisheries data indicates at least 85 percent of most federally tracked species have shifted north and/or into deeper levels of the sea over the past half century, especially within the last ten or fifteen years.

For example, over the past decade, black sea bass have migrated up the East Coast into southern New England. They are now caught in traps once designated for lobsters.[55]

Lobsters love cool water, where the temperatures stay below 68 degrees Fahrenheit. In the fall of 1999, lobstermen in the Long Island Sound saw a vast lobster die-off after a long stretch of unusually high water temperatures.[56]

And this wasn't a singular event. There's been a steady decline of lobster life in areas south of Maine. From 1996 to 2014, New York's registered lobster landings dropped 97.7 percent.[57] The story is much the same in Connecticut, where landings fell 96.6 percent from the most profitable year, and in Rhode Island, which saw a 70.3 percent reduction.[58]

Where some regions saw losses, others made gains. For example, NOAA scientists point to Maine as the primary state for lobster exports today. But if you go back to the 1980s, there had previously been a thriving lobster industry in New York and southern New England. Those lobsters moved north and became "Mainers." While southern New England lobstermen have found increasingly empty traps since the mid-90s, Maine's lobster fishery has boomed. But will it stay that way?

There's evidence that sea surface temperatures in the Gulf of Maine are warming 99 percent faster than sea surface temperatures on the rest of the planet, and will rise by at least twenty degrees Fahrenheit by the end of the century.[59]

Maine's waters, NOAA scientists say, aren't guaranteed to remain friendly to lobster, which means one day the state's lobster industry may find itself in the same position as that of southern New England.[60]

Sustainable Seafood

In the United States, both wild-caught and farmed fish and shellfish are managed under a system of enforced environmentally responsible practices. Both wild-capture and farmed fish are essential for ensuring sustainable seafood supplies are available for our nation and the world.[61]

The Monterey Bay Aquarium Seafood Watch program creates science-based recommendations on ocean-friendly seafood choices.[62]

(Source: Monterey Bay Aquarium Foundation)

Seafood Watch Guide

How to Use This Guide	Best Choices
Most of our recommendations, including all eco-certifications, aren't on this guide. Be sure to check out SeafoodWatch.org for the full list. **Best Choices** Buy first; they're well managed and caught or farmed responsibly. **Good Alternatives** Buy, but be aware there are concerns with how they're caught, farmed, or managed. **Avoid** Take a pass on these for now; they're overfished, lack strong management, or are caught or farmed in ways that harm other marine life or the environment.	Abalone (farmed) Arctic Char (farmed) Barramundi (US & Vietnam farmed) Bass (US farmed) Catfish (US) Clams (farmed) Cockles Cod: Pacific (AK) Crab: King, Snow & Tanner (AK) Lionfish (US) Mussels (farmed) Oysters (Canada, US & farmed) Prawn (Canada & US) Rockfish (AK, CA, OR & WA) Sablefish/Black Cod (AK) Salmon (New Zealand) Sanddab (CA, OR & WA) Scallops (farmed) Shrimp (US farmed) Squid (California market) Sturgeon (US farmed) Tilapia (Canada, Ecuador, Peru & US) Trout (US farmed) Tuna: Albacore (trolls, pole and lines) Tuna: Skipjack (Pacific trolls, pole, and lines)
Good Alternatives	**Avoid**

Clams (US & Canada wild)
Cod: Atlantic (handlines, pole & lines)
Cod: Pacific (Canada & US)
Lobster: Spiny (Bahamas & US)
Mahi Mahi (Costa Rica, Ecuador,
    Panama & US longlines)
Monkfish (US)
Octopus (Canada, Portugal & Spain pots
    and traps, HI)
Oysters (US wild)
Pollock (Canada longlines, gillnets & US)
Salmon: Atlantic (BC & ME farmed)
Salmon (CA, OR & WA)
Scallops: Sea (wild)
Shrimp (Canada & US wild,
    Ecuador & Honduras farmed)
Squid: Jumbo
Swordfish (US)
Tilapia (Colombia, Honduras, Indonesia,
    Mexico & Taiwan)
Trout (Canada & Chile farmed)
Tuna: Albacore (US longlines)
Tuna: Skipjack (free school, imported
    trolls, pole and lines, US longlines)
Tuna: Yellowfin (free school, trolls,
    pole and lines, US longlines)
	Basa/Pangasius/Swai
	Bass: Striped (US gillnet, pound net)
	Cod: Atlantic (gillnet, longline, trawl)
	Cod: Pacific (Japan & Russia)
	Crab (Argentina, Asia & Russia)
	Halibut: Atlantic (wild)
	Lobster: Spiny (Belize, Brazil,
	    Honduras & Nicaragua)
	Mahi Mahi (Peru, Taiwan)
	Orange Roughy
	Octopus (other imported sources)
	Pollock (Canada trawls & Russia)
	Salmon (Canada Atlantic, Chile,
	    Norway & Scotland)
	Sharks
	Shrimp (other imported sources)
	Squid: Argentine shortfin, Indian, Japanese
	    flying, mitre & swordtip
	Swordfish (imported longlines)
	Tilapia (China)
	Tuna: Albacore (imported except trolls,
	    pole & lines)
	Tuna: Atlantic Bluefin (imported longlines)
	Tuna: Pacific & Southern Bluefin
	Tuna: Skipjack (imported purse seines)
	Tuna: Yellowfin (imported longlines except US)

Avocados

Leathery on the outside, yet green, and creamy on the inside—avocados have long been a food favorite. Today, the fruit (yes, it's a fruit!) has a thriving international following that's synonymous with millennial pop culture. Did you know avocados are one of the most popular and highly photographed foods on Instagram?[63]

For decades, California and Mexico have proved to be ideal spots to grow avocados. But an increasing number of heat waves and periods of drought have posed challenges to meet consumers' record demand. Prices for the green super-food have gone up as a result.

Successful avocado farms are dependent on a precious natural resource—water. It's hard to imagine, but each avocado requires 320 liters of water to grow.[64] These tasty fruits are also impacted by extreme weather. In California, NOAA scientists say increased tree water use due to dry Santa Ana wind events, sensitivity to warmer August temperatures, and wildfires could cause a 45 percent reduction in avocado yields statewide by 2060.[65]

Given the high profit to be made on the fruit, the cultivation of avocado is often prioritized above other crops. Since avocados tend to be grown for export and not for local communities, this practice hurts regional food security.[66]

According to the Rainforest Alliance, Mexican avocado farmers are under pressure to meet rising demand and often resort to expanding their cropland through deforestation. The Rainforest Alliance has sought to reduce environmental impacts by certifying 900 acres of avocado farms in Jalisco, Mexico—the first avocado certification of any kind in that country—"helping to minimize adverse environmental impacts and sustain the production of this delicious fruit."[67]

Chocolate

A tree of chocolate! Sounds like something out of a Willy Wonka movie.

Unfortunately, while cacao does grow on trees, candy-covered branches would not thrive in the continental United States' temperate climate, according to climate.gov. Chocolate grows best in the tropical regions—where it's so hot it would melt in your hands.

Ideal weather cacao growing conditions feature warm temperatures, high humidity, abundant rain, and nutrient-rich soil protected from the wind. Over the next several decades, though, those perfect chocolate-favored spots may lose their valued climatic properties.

Chocolate is typically grown within 10 degrees north and south of the equator. The world's leading producers are Indonesia, Côte d'Ivoire, and Ghana, with the latter two producing over half of the world's chocolate.[68]

Research highlighted in the Intergovernmental Panel on Climate Change (IPCC) report *Climate Change 2014: Impacts, Adaptation, and Vulnerability* indicate that, under a "business as usual scenario, those countries will experience a 3.8°F (2.1°C) increase in temperature by 2050 and a marked reduction in suitable cacao cultivation area. There's already a premium on suitable land for cacao. By 2050, rising temperatures are also projected to push cacao cultivation areas further uphill. The IPCC reported that Côte d'Ivoire and Ghana's optimal altitude for cacao cultivation is expected to rise from 350–800 feet to 1,500–1,600 feet above sea level.[69]

Keep in mind, as in many weather scenarios, it's not the heat; it's the humidity. The danger global warming brings to chocolate comes from an increase in *evapotranspiration*—water release from plant leaves. The warmer and dryer temperatures projected for West Africa by 2050 based on many current carbon dioxide emissions scenarios would be detrimental to cacao plants.[70]

Looking for Sustainable Chocolate?

Here are some organizations that offer directories and more information:

- Rainforest-alliance.org
- Fairtradecertified.org
- thegoodtrade.com
- Foodispower.org
- leafscore.com

Chocolate isn't likely to go extinct, though. (Whew!) For example, researchers in Brazil are using innovative processes to cultivate new varieties of chocolate from the Amazon's native cacao trees and other species like the cupuaçu, a tree fruit often made into a sour juice but with a pod similar to cocoa.[71]

If you are looking for chocolate grown sustainably, check the packaging for fair-trade sourcing. The Rainforest Alliance certified chocolates (and other items) with a seal meaning that the product is eco-friendly and produced ethically by farmers, foresters, and corporations. The incentive is for companies and sellers to work together to create a world where people and nature thrive in harmony."[72]

Farmers' Markets

There are more than 8,500 farmers' markets across the US.[73] They provide healthy, sustainable food options. To find one where you are located, check the directory at https://www.farmersmarket.net.

Behind each concession stand of sweet peaches and strawberries, crisp lettuce, crunchy carrots, and freshly hatched eggs, there is a story. One of those tilling tales is found in Georgia, where fourth-generation farm owner Mary Blackmon made an unexpected return to the family business.

"I never lived on a farm until my mid-forties," she said. "My grandparents' farm was on the border of Arkansas and Louisiana. In 2008, my uncle called and said it would be sold since now there'd be no one left to run it. I decided to run it myself. It was painful to let such heritage go."

Mary later launched a new website: farmstarliving.com, focusing on the lives of farmers, farm-fresh food, and wellness. It's a great guide to what's in season, where to find it, and nature-inspired health and beauty tips.

Over the past decade, Mary's written about or worked with hundreds of farmers across the country and shares perspectives on growing food sustainably.

Don't Throw It Away!

The Food and Agriculture Organization (FAO) estimates that approximately one-third of all food produced for human consumption in the world is lost or wasted each year.[74]

Interview with Mary Blackmon,
Founder, Farm Star Living

What are the trends you're seeing in sustainable farming?
It's a two-part answer. Farmers have always been mindful of the land because they are stewards of their land. They've always had environmental concerns in issues close to the heart. Moving forward, the second part is the consumers. Their interests reflect more environmentally conscious approaches to their food. Consumers are on that same page, but they're also asking the farmers to rise in areas that they fall short in. Many traditional farming companies have now embraced organic farming. Keep in mind, it can take a while to convert a conventional farm to organic methods.

How are farmers implementing these practices?
Farmers are now looking at more ways to conserve water and use fewer natural resources. In California, drought has required them to be creative and innovative in finding ways to recycle water. They're implementing better solar power systems and wind turbines using all of these new efficiencies to conserve our natural resources. For example, there's a farm called Limoneira in California. They're citrus growers in Ventura that have been around for over a century. They've got a 10-acre facility on-site to receive green materials (lawn clipping, leaves, bark, plant materials) from throughout Ventura County. The material is converted into mulch that is spread in their lemon orchards to curb erosion, improve water efficiency, reduce weeds, and moderate soil temperatures.

How have more frequent extreme weather events impacted the farmers you know?

I wanted to showcase and celebrate farmers because many people do not understand who they are, what they do, and why they do it. It's their passion; it's in their blood; it's not about really earning a dollar. You don't always make money; you often lose money for years at a time. A big part of that uncertainty is the weather.

I can speak from personal experience in three of the last four years. We've had such massive rainfalls in the spring. We've had to delay our planting, or it's stunted. It's made the whole farming process much more challenging. We've had a hard time at harvest because the winds would blow everything down. You can't capture the food in the rain—it was a scramble.

And I work with farms all over the country. For example, on the West Coast in the states of Washington and Oregon, apple farms had completely unexpected frost, and the townspeople had to gather together to salvage the fruit. I believe they picked over one hundred thousand apples in a matter of a week before the ice came in; they picked them all by hand. I also work with many farms in Nebraska and Idaho, and they had such severe floods, which significantly impacted the potato industry. Hurricanes on the East Coast have damaged watermelon crops or their equipment.

"Food waste contributes to greenhouse gas as it decomposes," said Dr. Gabrielle Lyon, a medical doctor who regularly speaks on the topic of food, health, and medicine. In the United States, 30 percent of all food, worth $48.3 billion, is thrown away each year. The EPA estimates that in 2018, more food reached landfills and combustion facilities than any other single material in our everyday trash. Food waste is responsible for 11 percent of global greenhouse gas emissions.[75]

And the gas food waste generates is potent. A *Washington Post* article, "Food Waste Is a Vastly Overlooked Driver of Climate Change," points out that decomposing food waste in landfills releases methane, a greenhouse gas that is at least twenty-eight times more potent than carbon dioxide."[76]

Food waste is also expensive and a huge contributor to water loss. According to a 2019 IPCC report, food waste costs about $1 trillion per year.[77] It is estimated that about half of the water used to produce this food also goes to waste. Agriculture is the most prominent human use of water.[78]

Yale Climate Connections reports food waste is a global problem that can also pose a paradox: "In developed countries, for instance, consumers, sometimes seemingly with abandon, simply discard what they see as 'excess' or 'surplus' food. In developing countries, much of the waste is brought about by a lack of refrigeration as products go bad between producers and consumers. Meanwhile, some two billion humans worldwide are overweight or obese even as nearly one billion are undernourished, highlighting the inefficiencies and inequities in food distribution."[79]

Food Waste Solutions

Wasted food is the single largest category of material placed in municipal landfills and represents nourishment that could have helped feed families in need. The EPA is taking action to fight food waste with its "US 2030 Food Loss and Waste Reduction" goal to help feed the hungry, save money for families and businesses and protect the environment. Led by USDA and EPA, the federal government seeks to work with communities, organizations, and businesses along with state partners and tribal and local governments to reduce food loss and waste by 50 percent over the next decade.[80] For more information, visit https://www.epa.gov/sustainable-management-food.

Individual Steps You Can Take to Reduce Food Waste

At the grocery store or when eating out:[81]

- Preplan and write your shopping list before going to the grocery store. Check your fridge to see what items you already have.
- Buy only what you need and stick to your shopping list. Be careful when buying in bulk, especially with items that have a limited shelf life.
- If available, purchase "ugly" fruits or vegetables that often get left behind at the grocery store but are safe to eat. "Ugly" fruits and vegetables are safe and nutritious and can sometimes be found at discounted prices.
- When eating out, ask for smaller portions to prevent plate waste and keep you from overeating.

Storage and prep in the kitchen:

- Check the temperature setting of your fridge. Keep the temperature at 40° F or below to keep foods safe. The temperature of your freezer should be 0° F.
- Refrigerate peeled or cut veggies for freshness and to keep them from going bad.
- Use your freezer! Freezing is a great way to store most foods to keep them from going bad until you are ready to eat them.
- Create a designated space in your fridge for foods that you think will be going bad within a few days.
- Consider donating your extra supply of packaged foods to a local food pantry or a food drive.
- Learn about food product dating. Many consumers misunderstand the purpose and meaning of the date labels

that often appear on packaged foods. Confusion over
date labeling accounts for an estimated 20 percent of
consumer food waste.

- Except for infant formula, manufacturers are not required by federal law or regulation to place quality-
based date labels on packaged food.
- Consumers should examine foods that are past their
"Best if used by" date for signs of spoilage.
- Consumers may want to avoid eating products that have
changed noticeably in color, consistency, or texture.
- Manufacturers apply date labels at their discretion and
inform consumers and retailers of the date they can expect the food to retain its desired quality and flavor.

Cooking, serving, and enjoying food with family and friends at home:

- Create new dishes and snacks with leftovers or items
you think will go bad if not eaten soon. Have a cook-off
to find out who can come up with the best dish.
- Follow the two-hour rule: For safety reasons, don't leave
perishables out at room temperature for more than two
hours unless you're keeping them hot or cold.
- If the temperature is above 90°F, food shouldn't be left
out for more than one hour. Also, remember to refrigerate leftovers within two hours.

10.

SUSTAINABLE FITNESS

Featuring a 1.5 km swim in the Hudson River, a 40 km bike ride along the Westside Highway, and a 10 km run in Central Park, the highly anticipated New York City Triathlon was gearing up for over 4,000 athletes to participate in its July 2019 competition. The summer event was carefully planned. Its seasonal timing offered advantages like lighter city traffic for running and cycling and smoother river conditions for swimming. But that year, though, one external factor overtook the excitement, coordination, and race anticipation: the weather was just *too hot*.

"After weighing all options to ensure the safety of athletes, volunteers, spectators, and staff due to the oppressive heat and humidity forecast on race weekend," organizers announced, "it is with great disappointment that we announce the cancellation of the 2019 Verizon New York City Triathlon."[1]

That same year, around the world, in Tokyo, athletes faced similar consequences at a 2019 training event for the women's triathlon in the 2020 Summer Olympics. Due to extreme heat, the International Triathlon Union cut the running portion in half.[2]

July 2019 wasn't just sweltering for sports events; it was the hottest month ever recorded on the planet. Even the Arctic felt the heat. Average Arctic sea ice set a record low for July, running almost 20 percent below average—surpassing the previous historic low of July 2012. The Arctic, scientists say, is warming at a pace twice as fast as the rest of the planet.[3] It's hampering animal athletes as well. The mushers of Alaska's famous Iditarod ended their races preemptively during a recent competition due to unseasonably warm temperatures and slushy conditions. It was the second-highest scratch rate in the competition's forty-seven-year history.[4]

From the North Pole down to the Southern Hemisphere, another environmental problem was emerging for outdoor athletes: Australian bush fire smoke. Air so thick it was hard for pro-golfers to see the fairway. The *Times of Malta* reported athletes complained of stinging eyes and trouble breathing during golf's Australian Open. "Raging fires across New South Wales wafted into Sydney. New Zealand's Ryan Chisnall, who is asthmatic, played in a mask."[5]

Climate change's impact of extreme heat and poor air quality can potentially impact anyone who exercises outdoors, even if you're solely a weekend warrior. There are ways to identify hazards and take precautions to reduce risk before you head outdoors.

Following the pandemic shutdowns of 2020, the quest to stay fit has sparked innovations in exercise trends. Sustainable advances are expressed through eco-friendly athletic wear, efforts to recycle sports gear, and by using fitness equipment that can actually generate its own power.

Stay at Home/Stay Fit

The COVID-19 pandemic of 2020 forced many to explore fitness options available outside of the traditional gym—due to temporary closings enforced by stay-at-home orders. This led to an increased demand for new ways to work out. For many, that was redesigning their home space into an exercise studio. Online marketplace eBay reported that between March and April 2020, sales of fitness equipment increased up to twenty times in some categories compared to the same timeframe in 2019.[6]

Connecticut-based Nicole Glor, the creator of the Nikki Fitness exercise videos and author of *The Slimnastics Workout*, says her clients embrace the concept of at-home exercise but also like to participate in the outdoor group classes she conducts remotely. "I have a class that I'll teach that is a walking workout with weights. This is something that you could do with your friend from Miami and your friend from Chicago. Everyone can see each other on this Zoom kind of screen. So that does save energy (you're not driving to the gym), and you have that camaraderie and support with friends. It brings the social aspect, just through technology," Nicole said.

During the 2020 pandemic, exercise classes popped up in local parks while keeping in mind COVID-19 restrictions and government ordinances. Gym owners found creative places to teach outside, from parking lots to unused sports stadiums. Group classes were reduced and reconfigured to more scenic spaces.

Social trends reflected these fitness endeavors. The term *outdoor gym* was searched for approximately 322 times in December 2020, according to Moz .com.[7] Google Trends indicates that this is 722 percent higher than this time five years ago.

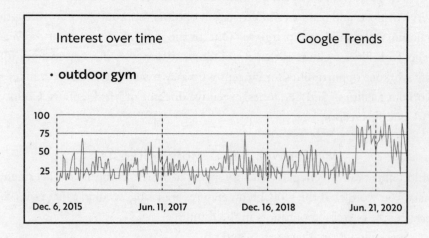

Free Fitness Outdoors

Other community-wide outdoor facilities strive to inspire fitness and be free of charge. The city of Toronto's new "fitness pod" in the community in Scarborough Southwest is designed as a circuit with three stations that can accommodate twenty-four different exercises, from beginner to advanced fitness levels. These no-cost equipment components are easily adaptable and have no moving parts, which increases their overall safety and minimizes the need for frequent maintenance.[8]

The concept was created by the San Francisco–based National Fitness Campaign, a consulting firm that partners with cities, schools, corporations, and design firms to fund and build outdoor fitness courts to improve Americans' quality of life. With over two hundred fitness courts open today, they seek to expand upon and optimize existing pedestrian infrastructure. Their futuristic-looking setup is designed to provide a seven-movement workout described on their website: core, squat, push, lunge, pull, agility, and bend.[9]

In October 2020, the National Fitness Campaign partnered with the NFL Panthers and Lowe's, Mecklenburg County, and Veterans Bridge, creating a nine-hundred-square-foot synthetic turf space for stretching, yoga, and other active workouts in Charlotte, North Carolina's Veterans Park.

The project is part of the Panthers and Lowe's collective military Salute to Service efforts and features a multi-station outdoor workout space for strength training, agility, and cardio fitness. "One of the important aspects of the Veterans Park Fitness Court is that it's outdoors. It provides the opportunity for safe workout opportunities for citizens who may have limited access to indoor workout facilities," said Lee Jones, executive director of Mecklenburg County Park and Recreation.[10]

Warmer Weather . . . More Workouts?

Mild temperatures generally tend to appeal to fitness enthusiasts. This might make you wonder: if the global temperatures are rising, wouldn't that point to *more* people exercising outdoors in the future?

Scientific research says yes . . . and no.

Dr. Nick Obradovich, PhD, who focuses on the social impacts of climate change at both Harvard's Kennedy School of Government and MIT, evaluated data from 1.9 million survey respondents with historical estimates of multiple climate models to project how future change may affect physical activity.

His findings, reported in the journal *Nature Human Behavior*, indicated that Americans will indeed be more likely to get outdoors, increasing their physical activity by as much as 2.5 percent by the end of the century. Now, that means more physical activity *overall* across the entire US; but how this actually plays out depends on where you live. Obradovich concluded, "Activity may increase most during the winter in northern states and decline most during the summer in southern states."[11]

Here's where part two of the answer to global warming and its impact on outdoor exercise comes into play. The study also found that weather extremes—like extremely hot or very cold days and rainy conditions—pointed to a *reduction* in physical activity. Looks like most people prefer "perfect weather" for their workouts.

Hot Workouts

Breathing harder as you run, jump, swish, or swoop is even tougher in extreme heat. Exercising in super-hot weather places additional strain on your body—which is already under exertion as you work out. Greater speed and higher intensity force your lungs to take in oxygen faster to provide adequate energy to your muscles. Your lungs must also work quickly and efficiently to remove carbon dioxide as you exhale. It's a process both on the field and inside your body. You may need up to fifteen times more oxygen during exercise.[12]

Ever get that feeling that you can't breathe any harder while sprinting across a finish line or during high intensity interval training (HIIT)? *VO max* is the maximum volume (V) of oxygen (O_2) your body can process when exercising. Greater aerobic fitness is indicated by higher VO max levels.[13] The diaphragm and intercostal muscles (located between the ribs) increase or decrease their pressure to cause air to either rush in or be forced out of the lungs.[14]

The higher the VO max, the more fit a person is.[15] In extreme heat and high humidity, dehydration causes a decrease in VO max, which means the body can't utilize oxygen efficiently to provide energy.[16]

According to the Cleveland Clinic, for every degree the body's internal temperature rises, the heart beats about ten beats per minute faster.[17] To cope with this extra demand, you're sucking in about ten times as much air as you normally would while resting. Internal circulation also speeds up to deliver the oxygen to the muscles so that they can keep up the pace based on your level of exertion.[18]

Perspiration allows your body to release heat, but it puts an additional strain on your cardiovascular system through fluid loss. When you sweat, you excrete sodium, potassium, and other minerals that your body needs to contract muscles and transmit information via your nervous system.[19]

Little Athletes in Heat

Extreme heat and exercise can be a dangerous combination, with consequences ranging from exhaustion to death. Exercise-related heat exhaustion is an illness caused by getting too hot during physical exertion. Not drinking enough fluids during exercise can also cause dehydration. Together, say experts at Johns Hopkins Medicine, "these things can make you collapse."[20]

Children are greater at risk for heatstroke, as they dehydrate faster than adults and are less likely to communicate when they physically need to stop activity due to overheating. For these reasons, there's been a growing concern about hot weather and children's sports. Heat illnesses are a leading cause of death and disability in young athletes.[21]

Every year, some nine thousand high school athletes are treated for heat-related illnesses. Young men make up a third of all heat-related emergency room visits in the US. Football players are eleven times more likely to suffer exertional heat illnesses than players of all other high school sports combined.[22]

Emergency room visits for heat illnesses increased by 133 percent between 1997 and 2006. Almost half of these patients were children and adolescents.[23] In 2017, a breakthrough report from the *Medical Consortium on Climate and Health*, which represents about 50 percent of the country's physicians, explained multiple ways climate change is already affecting human health. Doctors from all over the country shared firsthand accounts of their experience with patients. For Alexandria, Virginia–based pediatrician Dr. Samantha Ahdoot, her contribution was even more personal—because the patient she described was her son: "My nine-year-old son Isaac was attending his last day of band camp when I received a call from the emergency room. He had collapsed in the heat and was rushed to the emergency room. . . . That day was part of a record-setting heat wave in Washington, DC, when the heat index reached over 120 degrees."[24]

Dr. Ahdoot's son survived, but she says the incident shows how extreme heat events, which are projected to increase due to climate change, pose serious hazards to children. "I believe pediatricians on the front lines of this urgent problem must speak out for children on issues that will harm the health and prosperity of our youngest generations."[25]

Dr. Susan Yeargin, associate professor of physical education and athletic training at the University of South Carolina, is an expert on thermoregulation, hydration, and external heat illness in active populations like young people playing sports. Her research looks for new ways to recognize and treat heat illnesses to prevent them from happening in the first place—especially in children. Dr. Yeargin says that over the past few decades, the science behind how children's bodies handle heat has been revisited.

"It used to be that we thought that part of the reason why the rate is higher was that people assumed that children had inferior internal cooling mechanisms versus adults. That's not true. They're not inferior. Children thermoregulate differently, yet equally as efficient as adults."

Dr. Yeargin says that initial laboratory studies looking at youth exercising in the heat, conducted in the 1970s and 1980s, reported that children had higher core temperatures, heart rates, and lower sweat rates than adults. The problem, according to Dr. Yeargin, was that many of these early studies compared children and adults completing treadmill exercise at the same absolute speed instead of having the two populations train at the same relative intensity.[26]

That's key to understanding the differences. Children have a lower perspiration rate than adults. This is because youths dissipate heat non-evaporatively; this biological function allows them to conserve more fluids. Adults drip more sweat than children, but sweat drippage does not dissipate heat.

When your sweat is unable to evaporate into the air—which is what happens in externally hot and humid environments—or it's not wicked away, then it's not actually cooling you down. In fact, it's not helping you at all. Your body temperature will continue to increase.

Dr. Yeargin points out that parents and coaches need to watch for warning signs for overheating in children as they may not speak up and remove themselves from the environment of extreme heat as readily as an adult would. Symptoms of heat exhaustion include:[27]

- Dehydration
- Headache
- Profuse sweating or pale skin
- Loss of coordination, dizziness, fainting
- Nausea, vomiting, diarrhea

- Persistent muscle cramps
- Stomach cramps

"As an adult, if we go outside and it's hot, and we're just recreational athletes, I would like to think that we would hopefully stop ourselves if we don't feel good. Whereas a child, if they're in a practice and a parent or coach doesn't recognize that it's dangerously hot outside, then the child may not remove themselves from that situation," Dr. Yeargin said.

A Degree of Timing

As the official liaison for the National Association of Athletic Trainers (NATA) to the American Red Cross, Dr. Yeargin says that mitigating heat-related illness can also come down to *when* you choose to work out. "I think youth also face higher rates of these incidents because of the time that they're practicing outdoors. Often, it's as soon as the parents can get off work—five or six o'clock. But in the summer, it's still a scorching time of day," Dr. Yeargin explained. This goes for adults as well. She advises people to avoid afternoon hours when high temperatures peak. For planning purposes, check the weather's hourly forecast as well as sunset times.

There is a crucial temperature threshold to be aware of before you exercise outdoors. "We know that when temperatures rise above 82 degrees, the likelihood for heat illness dramatically increases."

Dr. Yeargin mentions it's also important to keep intensity in mind. For runners, this means pacing yourself with the outdoor environment in mind. "If a twelve-minute mile is a low intensity for you, you may be able to go out there longer than someone who runs at a much faster pace. But keep in mind how long you're out there as well. After two hours of exercising in the heat, we see research on sports that there's a higher risk for heat illness."

Wildfire Smoke

Exercising outdoors amid wildfire smoke may be more harmful than other types of air pollution. One study found that wildfire smoke could be up to ten times more dangerous than other air pollution sources, such as vehicles or

Additional tips to keep in mind if you plan to exercise when it's hot outside:

- Wear and reapply sunscreen.
- Drink more water than usual, and don't wait until you're thirsty to drink more. Muscle cramping may be an early sign of heat-related illness.
- Monitor a teammate's condition and have someone do the same for you.
- Wear loose, lightweight, light-colored clothing.
- Don't take medications that can intensify symptoms of heat exhaustion. Both caffeine and alcohol can accelerate the effects of dehydration.
- Consider exercising with a workout buddy. Oftentimes, it's difficult to notice when your own body is overheating.

industry.[28] In an interview with the *New York Times*, Dr. Jennifer Stowell, a postdoctoral associate at Boston University's School of Public Health, explained the reason behind this might be due to the "distinct mix of particulates that activate inflammatory cells deep in the lungs while hindering other cells that can dampen the inflammatory response later."[29]

Fitness expert Nicole Glor says the subject of wildfire smoke as a hindrance has come up more frequently in her clients' concerns. "The big thing that I heard mentioned is smoky air—when there are forest fires, whether it's Colorado or Australia. Or the sky looks red from smoke particles in San Francisco. It is affecting people. Sadly, these fires are happening in the places where people want to be outdoors. They want to be enjoying wellness outside, and they're not able to for many months of the year due to poor air quality," Nicole said.

Even a short run outside could put your health at risk. One study from the University of British Columbia found that the dangerous particles in wildfire smoke can impact a person's health within an hour.[30]

Smoky air affects even the most elite athletes. At the Australian Open in 2020, bushfire smoke sent top women's tennis player Dalila Jakupovic into a "coughing fit that forced her to leave the court and withdraw from her qualifying match."[31] Male tennis star Novak Djokovic said in a CNN interview that a delay might be necessary to wait for a reprieve in the smoky air before play.[32]

During exercise athletes breathe harder and increase the amount of air they inhale into their lungs. And they're doing it at a faster pace. "Because they work so hard and breathe so much, athletes turn out to be a sensitive subgroup to pollutants," says Ed Avol, a professor of clinical preventive medicine and air pollution expert at the University of Southern California, told Americanupbeat.com.[33]

Air Quality Outdoors

When planning outdoor exercise, it's important to map out your route or destination in advance, keeping in mind that pollution levels are likely to be highest within a quarter-mile of a road. How can you decide whether the air quality is too dangerous to exercise outdoors?

The EPA's US Air Quality Index is a good place to start. It uses color-coded categories and provides specific information about air quality in your area, which groups of people may be affected, and steps you can take to reduce your exposure to air pollution. It's also used as the basis for air quality forecasts and reporting.

The EPA has issued a national index for air quality since 1976 to provide an easy-to-understand daily report on air quality in a format that's the same from state to state. It covers the five major pollutants that are regulated by the Clean Air Act: ozone, particle pollution (also called particulate matter), carbon monoxide, nitrogen dioxide, and sulfur dioxide. Each pollutant is generally based on the health-based national ambient air quality standard for that pollutant and the scientific information that supports that standard.

Metro areas with a population of more than 350,000 are required to report the daily air quality index. For more information, visit https://epa.gov/outdoor-air-quality-data and airnow.gov/aqi.

Air Quality Index Forecasts

Air quality predictions are usually issued in the afternoon for the following day by state and local forecasters across the country. The information is put together using several tools, including weather forecast models, satellite images, air monitoring data, and computer models. These are all used to estimate how pollution travels in the air. Forecasts for ozone and particle pollution are most commonly issued, as they are two of the most widespread pollutants in the US. Some additionally address nitrogen dioxide and carbon monoxide.

- For ozone, an AQI forecast focuses on when average eight-hour ozone concentrations are expected to be the highest.
- For particulate matter, predictions center on the average twenty-four-hour concentration for the next day.
- AQI forecasts the next day's air quality and which groups of people may be affected, and steps individuals can take to reduce air pollution exposure.
- Many also provide a "forecast discussion," which also lets you know if there are times when air quality is expected to be better.

These forecasts can change, so it's a good idea to check them every day.

The Energy of Exercise

Ever add up the number of calories you need to burn on a treadmill to make up for a slice of pizza? How about how many miles you would need to pedal on a bike to power up all the lights at your gym? Energy exerted working out can be measured not only by offsetting indulgences but also in watts, just like with a light bulb.

For example, Tour de France cyclists pedal up about 500 watts for hours on end, and can hit 1,500 watts in short bursts.[34] Throughout one hour of intense exercise, elite athletes can produce about six watts of energy per kilogram of body weight. Divide that output in half, and you have an average fit person generating about three watts/kg per hour.[35]

This workout-for-wattage concept inspired Adam Boesel, when he was a personal trainer in 2008, to attach a small wind turbine–like device to his bicycle so that it could function as a generator.[36] His ambitious goal was to create a gym whose electricity was generated entirely by the members' exercising.[37] "That didn't happen—not even close," he said. "But I did create a gym that generated 35 percent of its energy needs using mostly solar and some human power. Our three-thousand-square-foot gym was getting by on about twenty kilowatt-hours per day, about the same as the average US apartment."[38]

An extreme weather event was the impetus for further innovation. After Hurricane Sandy in 2012, Adam was contacted by a representative of the New York City mayor's office who was exploring ways to help people whose power was out to charge their cell phones. This spurred Adam to develop the UpCycle Ecocharger. It can turn almost any bicycle into a generator by swapping out the back wheel and putting it up on a stand.[39] Adam's website, greenmicrogym .com, reports that the Ecocharger has been sold worldwide and was featured at the 2015 Pan Am Games.[40]

Run faster. Pedal harder. Create more power! At the Eco Gym at the Imaginarium in Rochester, New York, green cardio machines motivate members to ramp up their intensity by showing them their power output in real time. Gymgoers can monitor their individual energy production on a large screen as they work out.

Millennials love this stuff, according to SportsArt, the largest manufacturer of eco-innovative fitness products. The Washington State–based company says their research indicates that both millennials and Gen Xers say that they would be more inclined to exercise regularly if their gym supported green initiatives. They also responded that the green fitness routine often inspires them to live more sustainably.[41]

Body-Powered

Engineers from the University of California San Diego have developed what they call a "wearable microgrid" that can power small electronics.[42] The innovative device harvests and stores energy directly from the human body. No bike or treadmill is required. It's all you.

Here's how it works: monitors are screen-printed on the fabric of a long-sleeved shirt with sensors that make contact with your forearms and sides near the waist. When you swing your arms while walking or running, the motion detected against your torso turns movement into electricity. Sweat is measured from other sensors near the nape of the neck, chest, and back, which gives the device another way to measure expended energy. Despite its being garnished with all that imprinted gadgetry, the shirt is flexible to wear and washable.

The dual technique of harvesting energy from different body parts through both arm movement and sweat also allows the shirt to power devices quickly and continuously. Engineers say it's a concept similar (on a much smaller scale) to the way a city's power grid integrates local, renewable power sources like wind and solar.

How was the wearable grid put to the energy test? After a monitored subject spent ten minutes on a bike or treadmill, followed by twenty minutes of resting, the system generated usable energy. It could power either a liquid crystal display wristwatch or a small electrochromic display—a device that changes color in response to an applied voltage—throughout each session.[43]

The researchers are working on other designs that may not even require vigorous exercise to generate power. The team is engineering ways to harvest energy even while the user is sitting inside an office, for example, or moving slowly outside.

Your Sneaker's Carbon Footprint

Every step you take on the road to fitness leaves some sort of a carbon footprint behind. Its size isn't determined by how fast you run or how big your feet are. Often, its impact is derived from the athletic shoes you wear. Your most comfortable sneakers, dependable tennis shoes, and other bouncy lace-ups—despite their plush cushioning—can clomp down hard on the environment.

MIT scientists say a typical pair of running shoes generates thirty pounds of carbon dioxide emissions.[44] How is that possible? Most of these pollutants are the byproduct of manufacturing and through acquiring and or extracting raw materials.

Sneakers are predominantly fabricated from plastic and plastic-like materials that are then molded and sewn together to fit your feet. Sounds simple, but the MIT researchers found that a typical pair comprises sixty-five distinct parts requiring more than 360 processing steps. It's not just molding and sewing; carbon dioxide is fed into the molten resin to create foam. (Think of all that cushioning that sneakers have.)

Assembling these small sneaker parts is an energy-intensive process. Much of a sneaker's carbon impact comes from powering manufacturing plants.

China is home to a significant portion of the world's shoe manufacturers, and coal is the dominant electricity source in that country.[45]

You're not the only one who loves their sneakers. Just like a runner racing to the finish line, the demand for athletic footwear isn't expected to slow down anytime soon. Current projections anticipate this market to exceed $95 billion globally by 2025.[46]

How to Recycle/Reuse Athletic Shoes

When cleaning out your closet, think twice before you toss sneakers in the trash. Disposed shoes can pile up in landfills and take up to forty years to fully decompose![47] Donating used items to charity organizations is a good way to get rid of what you no longer need and potentially help someone in the process.

Here are some organizations to donate or recycle athletic shoes:

- **More Foundation**—The MORE Foundation Group has been diligently collecting your gently used athletic shoes for over ten years with over eight hundred drop-off locations in the United States. (Morefoundationgroup.org)
- **Soles4Souls**—Soles4Souls has collected more than 30 million pairs of shoes since 2006 and distributed them to children in need from 127 countries. Soles4Souls accepts all types of shoes, even flip-flops and dance shoes if they are new or gently worn.[48] (Soles4Souls.org)
- **Hope Runs**—Hope Runs is a nonprofit group working in Kenya and Tanzania, using athletics, education, and social entrepreneurship to help locals in need. They accept donations, including running shoes.[49] (HopeRuns.org)
- **One World Running**—a group of runners in Boulder, Colorado, collects, washes, and sends new and "near-new" athletic shoes and other athletic equipment to Third World countries. (Oneworldrunning.com)
- **Nike Reuse-a-Shoe**—Grinds your old running shoes into material that makes athletics and playground surfaces. (nikereuseashoe.com)

Workout Wear

Yoga pants. Biking shorts. Jog bras. Tennis whites. Golf shirts. As with like fitness gear for your feet, the quest to make more sustainable workout wear for your body is a growing trend. The UK fashion site Lyst.com says searches for sustainable fashion on their website went up 75 percent from 2018 to 2019.[50] Customers wanted to know whether athletic wear was eco-friendly—using recycled plastic or regenerated nylon, organic cotton, or recycled polyester. Searches for specific sustainable materials commonly used in activewear, like econyl, repreve, tencel, and organic cotton, showed increases between 40 to 100 percent.[51]

The fashion industry has a sizable carbon footprint. According to the World Bank, it's responsible for 10 percent of annual global carbon emissions. That's more than all international flights and maritime shipping combined. At this pace, the fashion industry's greenhouse gas emissions will surge more than 50 percent by 2030.[52]

Shopping for new outfits in general is also on the rise. The average consumer buys 60 percent more clothing items than they did fifteen years ago.[53] Take a look in your closet. How many things have you kept for years, and how much is new? A National Association of Professional Organizers study estimates 80 percent of your clothing is worn only 20 percent of the time.[54]

Conserving natural resources is a priority for green living. The fashion industry is the second-largest consumer of water worldwide, generating around 20 percent of wastewater and releasing half a million tons of synthetic microfibers into the ocean each year.[55] Much of this comes from manufacturing and the lifecycle of apparel in general. As explained in a 2018 Quantis study, a lot more goes into making clothes than just trying them on. These energy-intensive processes include:[56]

- **Fiber Production**—the extraction and processing of fibers.
- **Transportation** from raw material extraction location and between the processing and the sales point.
- **Yarn Preparation**—transportation and spinning of yarn from both filament and staple fibers.

- **Fabric Preparation**—knitting and weaving yarn into fabric.
- **Dyeing and Finishing**—bleaching and dyeing fabric finishing.
- **Assembly**—cutting and sewing of fabric into apparel products.
- **Packaging Production**—including raw material extraction for secondary packaging, including cardboard use.
- **Distribution**—transportation from assembly location to retail stores or customers.
- **End of life**—processes involve collecting and managing apparel products at the end of their useful life (incineration and landfilling).

All of that adds up! Which is all the more reason to look for outfits that will last beyond one season. For athletic wear, this means choosing comfortable and sustainable fabrics that wash well. There are many available options. Cotton may come to mind first. It's certainly a popular pick for T-shirts and socks. This fabric has been around for thousands of years. It's drought- and heat-tolerant. Technological innovations have allowed cotton to become less water reliant than it used to be. Organic cotton, according to Eco-stylist.com, is even better. It's grown "without all the harmful pesticides and produced without the dangerous chemicals that normal cotton uses."[57] Certified organic cotton is a much cleaner alternative, according to totebagfactory.com, as it manages to keep water, soil, and air clean by not using harmful pesticides, herbicides, fertilizers, or genetically modified crops. It also releases 46 percent less CO_2 during production, which means that it also saves energy.[58] Recycled cotton is even more sustainable than organic. It uses even less water and requires less energy expenditure throughout its life cycle. Most of this recycled fabric is made from cotton scraps produced by sewing plants worldwide that would otherwise have ended up in a landfill.[59]

Hemp, one of the oldest known fibers, dates back to 8000 BCE. The fabric requires even less land and water use than cotton. It's made with fiber from the stalks of the *Cannabis sativa* plant. It's versatile and enriches the soil through aeration, which adds oxygen and promotes bacterial growth. Hemp is grown without herbicides, pesticides, or chemical fertilizers and is biodegradable.[60]

Bamboo, like hemp, requires no pesticides and improves soil quality. Bamboo grass grows fast and can be harvested after two or three years. This resilient plant can thrive in rocky or sandy soil. Bamboo fabric is so soft that it's even used for baby clothing. It also absorbs moisture, is hypoallergenic, and is antibacterial.[61]

Citrus Twist

Sicilian start-up Orange Fiber solves two sustainability issues at once. It uses food waste to produce sustainable fashion.[62] In Italian, *pastazzo* is the cellulose and residue left behind after citrus juice is industrially produced. Each year about one million tons of citrus fruit pastazzo is a leftover by-product disposed of as waste. Orange Fiber found a fruitful solution.

The Catania-based company created patented material from citrus pastazzo. Their fabrics are formed from a silk-like cellulose yarn that can blend with other materials. The resulting 100 percent citrus textile features a soft and silky hand-feel and is lightweight and opaque or shiny when used in its purest form. The fashion blog unnatisilks.com says the fabric is cuddly, strong, and drapes beautifully, and it's anti-wrinkle.[63] Aside from looking pretty, the orange yarn has an additional benefit: thanks to nanotechnology, the material still contains the essential oils and vitamin C that are present in the citrus fruit peel.[64] According to Orange Fiber, the skin can absorb these oils (even though the fabric does not feel greasy). The oils are guaranteed to last at least twenty washing cycles, but the company is experimenting with recharging methods with special fabric softeners.[65]

The start-up's vegan silk was first brought to market by Salvatore Ferragamo in a 2017 capsule collection on Earth Day. In October 2020, the company completed creating a new plant in Sicily and produced the first ton of new sustainable fiber.

Refresh Responsibly

Speaking of landfills and decomposition times, those plastic water bottles that provide vital hydration to your fitness routine can take *hundreds of years* to degrade in a landfill. Plus, they accumulate 78 million tons of plastic packaging, about 32 percent of which is not recycled. This means that when you toss your empty bottle in a bin rather than reusing it to carry more water, it often ends up floating in it—and polluting the oceans.[66]

One sustainable solution is to switch from plastic to a reusable BPA-free eco-friendly water bottle for your workout. This small effort could actually make a big difference. Americans purchase about 50 billion water bottles per year, averaging about thirteen bottles per month for every person in the US. By using a reusable water bottle, you could save an average of 156 plastic bottles annually.[67]

When your workout is complete, there's another sustainable action to consider that also involves water: taking shorter showers. Water is a precious natural resource that you may not realize is being wasted down your drain. According to Boston University researchers, an average shower uses about five gallons of water per minute. If you shorten your shower by two minutes, you can cut your water use by ten gallons.[68]

Sustainable fitness is becoming more achievable through small steps you can take that may help raise your athletic game while lowering your carbon footprint.

CONCLUSION

As humans, our perception of risk is based on our assessment of the probability, propensity, and severity that each potential harm may bring. Our brains are wired to react quickly when under duress. When we perceive an obvious threat—one that we know might hurt us—our mind activates our motor functions to allow us to flee from peril quickly.

There's a long biological history behind this process. "Our brains still resemble those of our ancestors, whose objective was to hunt, scavenge, and hopefully stay alive until age thirty," says cognitive behavioral scientist Dr. Sweta Chakraborty, PhD.

In modern times, things have gotten more complicated when it comes to risk evaluation. Today's environment can be overwhelming and is filled with problems we have no prior experience figuring out. "Our brains haven't evolved nearly as fast as the risk landscape around us. We're still operating cognitively as though our risks are immediate and we have to make decisions quickly," says Dr. Chakraborty, who specializes in conveying credible science information. She says that one of the challenges of explaining climate change's ill effects on the general public is that the threats seem far off rather than right in front of us.

"The ripple effects of climate change tend to be slow-moving and more likely to affect our grandchildren, maybe great-grandchildren. Our brain doesn't register what should be a much more rational response to preventing some of these risk scenarios," she said.

Scientists who study climate change say new information is constantly unfolding. "When we see how things play out, these changes are happening much faster than our scientific models," says Dr. Drew Shindell, Distinguished Professor of Earth Science of Duke University who's previously testified on climate issues before both houses of the US Congress (at the request of both parties) and developed a climate-change course with the American Museum of Natural History. Dr. Shindell points to examples such as the rapid decline of Arctic sea ice, the frequency of the powerful tropical cyclones, sea-level rise, and extreme heat.

Medical experts agree that while these atmospheric changes are happening all around us, they can seem far off and faceless. "If only climate change had a villain. If only it was clear, this blatant transgression. We would be much more ready to do something about it," said Dr. Aaron Bernstein, the interim director of the Center for Climate, Health, and the Global Environment at Harvard's T. H. Chan School of Public Health.

This doesn't mean that fear should take over as we identify and confront these environmental risks to our well-being. Instead, we can pay more attention, gather greater knowledge, and take precautions to protect ourselves and those we care about. These proactive measures come in many forms: monitoring those most vulnerable. Staying on alert for severe weather. Enacting safeguards to prevent becoming infected with vector- or waterborne diseases. Recognizing the serious symptoms of seasonal depression or post-traumatic stress following natural disasters by prioritizing mental health.

Global warming makes us consider old foes like allergies in a new light. Fewer frost-free nights and more carbon dioxide in the air have led to more robust and longer allergy seasons—powerful enough to incite new allergies for those who previously never had them. Awareness of this longer and stronger season and reducing the chances of exposure to allergens both inside and outside the home are steps you can take to breathe easier.

Temperature, sunlight, and humidity levels can factor into feelings of depression and anxiety and trigger flare-ups of certain autoimmune diseases, like lupus, arthritis, and psoriasis. Detecting and avoiding (as much as possible) weather-related triggers are important parts of managing the chronic challenges of these ailments.

Wildlife and nature are intertwined with many aspects of human health. This delicate balance, when offset by deforestation, land development, and even ecotourism, can have dire consequences. Human interaction with wildlife can contribute to the spread of viruses. Geographical risk areas for vector and waterborne diseases are changing. This means Lyme disease, West Nile and Zika virus, and vibrio bacterias in water can pose new dangers in previously unlikely areas. As this book outlined, recognizing the risk and signs of these diseases early is crucial for effective treatment.

As ominous as flesh-eating bacteria sounds, anxiety has a way of digging under one's skin too. A very real casualty of climate change that this book addressed is young people's fear for the future of this planet. Generation Z and millennials, in particular, are tortured by this worry, known as eco-anxiety. The preceding pages shared the ways in which people of all ages are dealing with their own climate grief and expert strategies for talking to children about their environmental concerns.

Conversations about meat-inclusive or plant-based diets can be polarizing. But this book set out to explain varied experts' points of view—specifically in relation to climate change—to provide a well-rounded perspective. Innovations in agriculture to reduce greenhouse gasses are at the forefront of farming research. Improvements in water and soil management and carbon capturing are helping ranchers find practical solutions.

Sweltering summer days are exacerbated in cities by the urban heat island effect. Excessive heat can not only be physically debilitating but also rob you of restorative sleep at night, which can reduce your daytime mental sharpness. Increasing tree canopy coverage and green roof technology are useful strategies to help reduce the ill effects of city heat.

Hot temperatures and poor air quality can pose challenges to our quest for physical fitness. Negative environmental impacts can be experienced by outdoor athletes of all levels, from elite Olympians to school-age sports teams. Avoiding intense physical activity on ozone alert days and during the hottest periods of the afternoon can help mitigate harm.

As a meteorologist, I've often talked about how important preparedness is in advance of extreme weather. This includes creating a family disaster plan, securing emergency supplies in advance, and following safety precautions before, during, and after a natural disaster. These actions are even more necessary as intense hurricanes, catastrophic flooding, destructive wildfires, and debilitating heat waves happen more often, and in places where they were previously less likely to occur.

There can be increased feelings of anxiety as a storm approaches, but for many, this mental anguish remains long after skies have cleared. This post-traumatic stress is magnified for those who are displaced from their homes, particularly young children. It's important to recognize the signs of PTS and PTSD. Symptoms, as shown in this book, can be addressed in a variety of ways, from professional therapy to spending healing time in nature.

Throughout the preceding chapters, the notion of *equity* repeatedly emerged in evaluating those disproportionately affected by climate change. Excessive heat, found in the hottest parts of cities, is often aligned with where the most impoverished families live. Communities of color are also more likely to be exposed to polluted air. African American boys are at greater risk to develop asthma.

When it comes to autoimmune diseases, women of color are two to three times more at risk for lupus than Caucasians. Low-income renters are particularly vulnerable to being displaced by natural disasters. Extreme weather events that hamper farming and food production greatly impact low-income Latinos, as many rely on the agricultural sector for their livelihoods. As the United States becomes increasingly diverse, understanding how climate change impacts people of different racial and ethnic backgrounds is imperative.[1]

The elderly, especially those who live alone, often face severe health consequences of excessive heat. But sweltering temperatures and unhealthy air quality can also pose problems for the young. For example, children are at greater risk of succumbing to heat-related illness while playing sports on sweltering days. Doctors recognize that children's bodies and adults' bodies dissipate heat in different ways, and children are less likely to realize when they're dehydrated. This puts them in greater danger as climate scientists project that heat waves are likely to occur more often.

"People often ask me why I became a pediatrician," said Dr. Bernstein. "Pediatricians primarily are in the business of preventing harm to children. And it's tough to think about that without thinking about climate change."

The mental health effects on children are notable, including those of anxiety or depression. These feelings can result when children are confronted with the prospect of global warming and its potential consequences on their future. They can also be devastated emotionally after experiencing an extreme weather event firsthand. Parents can help by addressing children's fears directly and exploring ways for them to become more environmentally conscientious. Providing kids with even seemingly small opportunities to build inner resilience outside of times of crisis may help strengthen their coping skills for when life's inevitable challenges arise.

Throughout researching and writing this book, principally during the COVID-19 pandemic, I felt it was important to provide readers with helpful, science-based mental health self-care strategies. Various experts interviewed advocated practicing meditation, mindfulness, and gratitude as ways to cope with stress and anxiety. These experts extolled the benefits of these practices in helping to boost wellness in mind, body, and spirit.

For each chapter in *Taking the Heat* I canvassed weather, climate, health, and scientific research to not only help you identify harms but also provide you with valuable tools and takeaways.

Let's keep this conversation going. It would be a privilege to hear from you via email at Bonnie@weatherandwellness.com or through my social media accounts on Facebook, Twitter, Instagram, and TikTok; you can find me under Weather and Wellness and also my name. Please also visit my website, weatherandwellness.com.

Thank you for reading this book, and I wish you all the best in good health and great weather!

ACKNOWLEDGMENTS

There are many people and organizations I would like to thank for making this book possible. First and foremost, I feel grateful and privileged to all the people I interviewed who shared their incredible personal stories with me and my readers. Their courage to talk about their struggles in their own mental and physical health was inspiring, and I appreciate their perspectives included in this book. I want to thank the many physicians, scientists, medical and government officials, and others who helped me provide readers with their exceptional expertise and insight. Everyone I interviewed is renowned and respected in their individual fields. Their contributions to this book are highly valued.

I'm so pleased that Luba Ostashevsky, the editor of my first book, *Extreme Weather*, when she was at Palgrave Macmillan, later became my literary agent at Ayesha Pande Literary. Luba's intelligence and knowledge were invaluable to me throughout this process—from the book proposal through publication by Simon Element and Simon & Schuster. I appreciate Luba's guidance, input, and friendship.

At Simon & Schuster, I want to thank Emily Carleton, Natasha (Tasha) Yglesias, and Veronica (Ronnie) Alvarado, my talented and astute editors. It's been great to work with you all on this book. The keen eyes of the copy editors and the support of Simon & Schuster's sales, marketing, design, production, and social media teams are also much appreciated.

My television agent, Adam Leibner, has represented me for a long time. He's not only a talented agent, he's also a good friend. I also want to thank Adam as well as Miae Shin at the United Talent Agency.

My dream to write a book was one I've had since childhood. I'm fortunate that now I can say that I've done it twice, after a lot of hard work. Writing this book was a much more intensive endeavor than writing my first book, as the subject matter for *Taking the Heat* required a greater amount of research, data review, and analysis. It was also written during the challenging and uncertain time of the COVID-19 pandemic.

Through endless days of research, interviews, analysis, and writing, I found there was still more to learn about the critical topics of climate change, health, and wellness than I realized. This only inspired me more to work harder in order to share this vital information with you. My goal is to help everyone who reads this. Thank you for giving me that opportunity.

I'm grateful to my friends and family who offered their support, particularly since there were many days when I disappeared from contact as I was fully entrenched in this project and writing twenty-four-seven. Thank you to Mandy Carranza, Lisa Singer, Dylan Leder, Andrew Ehinger, Kathryn Prociv, Frances Rivera, and Abby Freed.

As I write this in New York City, I'm thinking about my family in Florida, whom I haven't seen in over a year due to the pandemic. Thank you to Roger, Andrea, Eric, and my adorable nephew, Adam, for all your support. My sister, Karyn, is always there with a check-in phone call or a funny story to make me smile. My mom, Judy, is full of praise, kindness, and encouragement. She teaches me to stay positive and persevere in tough times. My dad, Jerry, is someone I've looked up to since I was a child. Back then, he told me he would always be there for me, and he's never let me down. I'm fortunate to have such wonderful parents. Hopefully, I'll get to visit everyone in Florida soon!

Lastly, I would like to offer a special word of encouragement to any young woman reading this who is pursuing her goals and going through a difficult or challenging time: Remember that with every setback comes an opportunity to revise and rebuild. Rejection is also part of the process. Stay focused, seek mentors, embrace constructive criticism, and you will get there.

Bonnie Schneider

New York City
2021

NOTES

Introduction

1. "1,001 Blistering Future Summers," Climate Central, July 9, 2014, https://assets
.climatecentral.org/pdfs/CityFutureTemps-PressRelease.pdf.

2. "Medical Alert! Climate Change Is Harming Our Health," Medical Society Consortium
on Climate & Health, March 15, 2017, https://medsocietiesforclimatehealth.org/wp
-content/uploads/2017/03/medical_alert.pdf.

3. Ibid.

4. David Herring and Rebecca Lindsey, "Global Warming Frequently Asked
Questions," Climate.gov, October 29, 2020, https://www.climate.gov/news-features
/understanding-climate/global-warming-frequently-asked-questions.

5. "Special Report: Global Warming of 1.5 °C—Summary for Policymakers,"
Intergovernmental Panel on Climate Change, 2018, https://www.ipcc.ch/sr15
/chapter/spm/.

6. Ibid.

7. "2020 Was Earth's 2nd-Hottest Year, Just Behind 2016," National Oceanic and
Atmospheric Administration, January 14, 2020, https://www.noaa.gov/news/2020
-was-earth-s-2nd-hottest-year-just-behind-2016.

8. Brian Barnett and Amit Anand, "Climate Anxiety and Mental Illness," *Scientific
American*, October 10, 2020, https://www.scientificamerican.com/article/climate
-anxiety-and-mental-illness/.

9. Prathik Kini et al., "The Effects of Gratitude Expression on Neural Activity," *NeuroImage* 128 (March 2016): 1–10, https://www.sciencedirect.com/science/article/abs/pii/S1053811915011532.

Chapter One
Eco-Anxiety

1. Jenni Gritters, "What Is Eco-Anxiety?" *REI Co-op Journal*, December 12, 2019, https://www.rei.com/blog/news/what-is-eco-anxiety.

2. "Greta Thunberg: The Rebellion Has Begun," Medium, October 31, 2018, https://medium.com/wedonthavetime/the-rebellion-has-begun-d1bffe31d3b5.

3. Mike Pearl, "'Climate Despair' Is Making People Give Up on Life," *Vice*, July 11, 2019, https://www.vice.com/en/article/j5w374/climate-despair-is-making-people-give-up-on-life.

4. Susan Clayton Whitmore-Williams et al., "Mental Health and Our Changing Climate: Impacts, Implications, and Guidance," American Psychological Association, March 2017, https://www.apa.org/news/press/releases/2017/03/mental-health-climate.pdf.

5. Abel Gustafson et al., "A Growing Majority of Americans Think Global Warming Is Happening and Are Worried," Yale Program on Climate Change Communication, February 21, 2019, https://climatecommunication.yale.edu/publications/a-growing-majority-of-americans-think-global-warming-is-happening-and-are-worried/.

6. "New APA Poll Reveals That Americans Are Increasingly Anxious about Climate Change's Impact on Planet, Mental Health," American Psychiatric Association, October 21, 2020, https://www.psychiatry.org/newsroom/news-releases/climate-poll-2020.

7. John Watson, "Climate Concern Fuels the Rise of 'Eco-Anxiety,'" *Medscape*, July 8, 2019, https://www.medscape.com/viewarticle/915145.

8. Marco Garcia, "The Most Jaw-Droppingly Beautiful Beaches You Can Visit in Hawaii," *Thrillist*, May 31, 2019, https://www.thrillist.com/lifestyle/honolulu/best-beaches-in-hawaii.

9. Craig K. Chandler, "How Family Size Shapes Your Carbon Footprint," Yale Climate Connections, March 29, 2019, https://yaleclimateconnections.org/2019/03/how-family-size-shapes-your-carbon-footprint.

10. Dani Blum, "How Climate Anxiety Is Shaping Family Planning," *New York Times*, January 27, 2020, https://www.nytimes.com/2020/04/15/parenting/climate-change -having-kids.html.

11. Katie O'Reilly, "To Have or Not to Have Children in the Age of Climate Change," *Sierra*, November 1, 2019, https://www.sierraclub.org/sierra/2019-6-november -december/feature/have-or-not-have-children-age-climate-change.

12. "California Statewide Fire Summary," CAL FIRE, October 30, 2017, https:// web.archive.org/web/20180602170815/http://calfire.ca.gov/communications /communications_StatewideFireSummary.

13. "Thomas Fire Information and Updates," County of Santa Barbara, https://www .countyofsb.org/thomasfire.sbc.

14. Christopher Weber and Daniel Dreifuss, "13 Dead in Southern California as Rain Triggers Mudslides," AP, January 9, 2018, https://apnews.com/60bdabd547a540b0b7 2da785739a9033.

15. Avichai Scher, "'Climate Grief': The Growing Emotional Toll of Climate Change," NBC News, December 24, 2018, https://www.nbcnews.com/health/mental-health /climate-grief-growing-emotional-toll-climate-change-n946751.

16. Eliza Barclay and Brian Resnick, "How Big Was the Global Climate Strike? 4 Million People, Activists Estimate," *Vox*, September 22, 2019, https://www.vox.com/energy -and-environment/2019/9/20/20876143/climate-strike-2019-september-20-crowd -estimate.

17. "Good Grief Network," Patreon, https://www.patreon.com/GoodGriefNetwork.

18. Daisy Simmons, "Climate Change Support Group Launches," Yale Climate Connections, November 28, 2016, https://yaleclimateconnections.org/2016/11/clim ate-change-support-group-launches.

19. Judy Wu, Gaelen Snell, and Hasina Samji, "Climate Anxiety in Young People: A Call to Action," *Lancet Planetary Health* 4, no. 10 (October 1, 2020): E435–E436, https:// www.thelancet.com/journals/lanplh/article/PIIS2542-5196(20)30223-0/fulltext.

20. Catherine Pearson, "How to Talk to Your Kids about Climate Change Without Giving Them Anxiety," *HuffPost*, November 4, 2019, https://www.huffpost.com

/entry/how-to-talk-to-your-kids-about-climate-change-without-giving-them
-anxiety_l_5db9d75de4b00d83f7218b16.

21. Ibid.

22. Mary Democker, "So Your Kids Are Stressed Out About the Climate Crisis: Here's
How to Help," *Sierra*, January 25, 2020, https://www.sierraclub.org/sierra/so-your
-kids-are-stressed-out-about-climate-crisis.

23. Ibid.

24. Susan S. Lang, "Camping, Hiking, and Fishing in the Wild as a Child Breeds
Respect for Environment in Adults, Study Finds," *Cornell Chronicle*, March 13, 2006,
https://news.cornell.edu/stories/2006/03/wild-nature-play-age-11-fosters-adult
-environmentalism.

25. "What Is Ecotherapy?" Earthbody Institute, https://www.theearthbodyinstitute.com
/ecotherapy.

26. "Ecotherapy/Nature Therapy," GoodTherapy, August 15, 2018, https://www
.goodtherapy.org/learn-about-therapy/types/econature-therapy.

27. MaryCarol R. Hunter, Brenda W. Gillespie, and Sophie Yu-Pu Chen, "Urban Nature
Experiences Reduce Stress in the Context of Daily Life Based on Salivary Biomarkers,"
Frontiers in Psychology, April 4, 2019, https://www.frontiersin.org/articles/10.3389
/fpsyg.2019.00722/full.

28. "MaryCarol Hunter: Feeling Stressed? Take a 'Nature Pill,'" Michigan News,
University of Michigan, April 8, 2020, https://news.umich.edu/feeling-stressed-take
-a-nature-pill-says-u-m-researcher/.

Chapter Two:
Natural Disasters, Trauma, and Gratitude

1. Christine Vestal, "'Katrina Brain': The Invisible Long-Term Toll of Megastorms,"
Politico, October 12, 2017, https://www.politico.com/agenda/story/2017/10/12
/psychological-toll-natural-disasters-000547.

2. Knvul Sheikh, "Natural Disasters Take a Toll on Mental Health," BrainFacts
.org, June 29, 2018, https://www.brainfacts.org/diseases-and-disorders/mental
-health/2018/natural-disasters-take-a-toll-on-mental-health-062818.

3. "How to Cope with Traumatic Stress," American Psychological Association, October 30, 2019, https://www.apa.org/topics/trauma/stress.

4. Barbara Brody, "How to Tell if You Have Normal Post-Traumatic Stress or Something More Serious," Health.com, October 31, 2018, https://www.health.com/condition /ptsd/ptsd-or-normal-post-traumatic-stress.

5. Matthew J. Friedman, MD, PhD, "PTSD History and Overview," National Center for PTSD, US Department of Veterans Affairs, https://www.ptsd.va.gov/professional /treat/essentials/history_ptsd.asp.

6. Jitender Sareen, "Posttraumatic Stress Disorder in Adults: Impact, Comorbidity, Risk Factors, and Treatment," *Canadian Journal of Psychiatry* 59, no 9 (September 2014): 460–67, https://www.ncbi.nlm.nih.gov/pmc/articles/PMC4168808.

7. Sri Warsini and Caryn West, "The Psychological Impact of Natural Disasters among Adult Survivors: An Integrative Review," *Issues in Mental Health Nursing* 35, no. 6 (June 2014): 420–36, https://www.researchgate.net/publication/264636089_The _Psychosocial_Impact_of_Natural_Disasters_among_Adult_Survivors_An_Integrative _Review.

8. Son Chae Kim et al., "Medium-Term Post-Katrina Health Sequelae among New Orleans Residents: Predictors of Poor Mental and Physical Health," *Journal of Clinical Nursing* 17, no. 17 (September 2008): 2335–42, https://pubmed.ncbi.nlm.nih .gov/18705709/.

9. Karen B. DeSalvo et al., "Symptoms of Posttraumatic Stress Disorder in a New Orleans Workforce Following Hurricane Katrina," *Journal of Urban Health* 84, no. 2 (March 2007): 142–52, https://www.ncbi.nlm.nih.gov/pmc/articles/PMC2231633/.

10. Y. Neria, A. Nandi, and S. Galea, "Post-Traumatic Stress Disorder Following Disasters: A Systematic Review," *Psychological Medicine* 38, no. 4 (April 2008): 467–80, https:// www.ncbi.nlm.nih.gov/pmc/articles/PMC4877688.

11. Orla T. Muldoon et al., "The Social Psychology of Responses to Trauma: Social Identity Pathways Associated with Divergent Traumatic Responses," *European Review of Social Psychology* 30, no. 1 (2019): 311–48, https://doi.org/10.1080/10463283.2020.1711628.

12. "What Is Posttraumatic Stress Disorder?" American Psychiatric Association, https:// www.psychiatry.org/patients-families/ptsd/what-is-ptsd.

13. "Post-Traumatic Stress Disorder," National Institute of Mental Health, 2020, https://www.nimh.nih.gov/health/publications/post-traumatic-stress-disorder-ptsd/index.shtml.

14. Dale Vernor, "PTSD Is More Likely in Women Than Men," National Alliance on Mental Illness, October 8, 2019, https://www.nami.org/Blogs/NAMI-Blog/October-2019/PTSD-is-More-Likely-in-Women-Than-Men.

15. Ju-Yeon Lee, Sung-Wan Kim, and Jae-Min Kim, "The Impact of Community Disaster Trauma: A Focus on Emerging Research of PTSD and Other Mental Health Outcomes," *Chonnam Medical Journal* 56, no. 2 (May 2020): 99–107, https://www.ncbi.nlm.nih.gov/pmc/articles/PMC7250671.

16. Dan W. Grupe and Jack B. Nitschke, "Uncertainty and Anticipation in Anxiety," *Nature Reviews Neuroscience* 14, no. 7 (July 2013): 488–501, https://www.ncbi.nlm.nih.gov/pmc/articles/PMC4276319.

17. Ibid.

18. "What Is Posttraumatic Stress Disorder?" PTSD Alliance, http://www.ptsdalliance.org/about-ptsd/.

19. Homepage, PTSD Alliance, http://www.ptsdalliance.org/.

20. "10 Unexpected Physical Symptoms of PTSD," PTSDUK, https://www.ptsduk.org/10-unexpected-physical-symptoms-of-ptsd/.

21. J. Douglas Bremner, MD, "Traumatic Stress: Effects on the Brain," *Dialogues in Clinical Neuroscience* 8, no. 4 (December 2006): 445–61, https://www.ncbi.nlm.nih.gov/pmc/articles/PMC3181836/.

22. Les Christie, "Which Natural Disaster Will Likely Destroy Your Home?" CNN Money, June 19, 2014, https://money.cnn.com/2014/06/19/pf/insurance/natural-disaster-risk/index.html.

23. Ibid.

24. "Billion-Dollar Weather and Climate Disasters: Overview," National Centers for Environmental Information, https://www.ncdc.noaa.gov/billions/.

25. J. T. Houghton et al., eds., "Climate Change 2001: The Scientific Basis," Intergovernmental Panel on Climate Change, 2001, https://www.ipcc.ch/site/assets/uploads
/2018/03/WGI_TAR_full_report.pdf.

26. "Global Warming and Hurricanes," Geophysical Fluid Dynamics Laboratory,
March 29, 2021, https://www.gfdl.noaa.gov/global-warming-and-hurricanes/.

27. Robinson Meyer, "What's Happening with the Relief Effort in Puerto Rico?"
Atlantic, October 4, 2017, https://www.theatlantic.com/science/archive/2017/10
/what-happened-in-puerto-rico-a-timeline-of-hurricane-maria/541956/.

28. Michon Scott, "Hurricane Maria's Devastation of Puerto Rico," Climate.gov,
August 1, 2018, https://www.climate.gov/news-features/understanding-climate
/hurricane-marias-devastation-puerto-rico.

29. Gary S. Votaw, "Tropical Storm Maria Floods Puerto Rico," Weather.gov, https://
www.weather.gov/media/sju/events/others/Maria.pdf.

30. Scott, "Hurricane Maria's Devastation of Puerto Rico."

31. Votaw, "Tropical Storm Maria Floods Puerto Rico."

32. Scott, "Hurricane Maria's Devastation of Puerto Rico."

33. "Save the Children Helping Children Fully Recover Two Years After Hurricane
Maria," Save the Children, September 20, 2019, https://www.savethechildren.org
/us/about-us/media-and-news/2019-press-releases/children-recover-two-years
-after-hurricane-maria.

34. Zara Abrams, "Puerto Rico, Two Years After Maria," *Monitor on Psychology* 50, no. 8
(September 2019): 28, https://www.apa.org/monitor/2019/09/puerto-rico.

35. Ibid.

36. "Greater Impact: How Disasters Affect People of Low Socioeconomic Status,"
Substance Abuse and Mental Health Services Administration, July 2017, https://
www.samhsa.gov/sites/default/files/dtac/srb-low-ses_2.pdf.

37. Abrams, "Puerto Rico, Two Years After Maria."

38. "Behavioral Health Conditions in Children and Youth Exposed to Natural Disasters," Substance Abuse and Mental Health Services Administration, September 2018, https://www.samhsa.gov/sites/default/files/srb-childrenyouth-8-22-18.pdf.

39. "How Extreme Weather Events Affect Mental Health," American Psychiatric Association, https://www.psychiatry.org/patients-families/climate-change-and-mental -health-connections/affects-on-mental-health.

40. Anna Merlan, "Is NY Prepared for Climate Change's Mental Health Crises?" *Village Voice*, October 11, 2016, https://www.villagevoice.com/2016/10/11/is-ny-prepared -for-climate-changes-mental-health-crises/.

41. J. Brian Houston et al., "2011 Joplin, Missouri, Tornado Experience, Mental Health Reactions, and Service Utilization: Cross-Sectional Assessments at Approximately 6 Months and 2.5 Years Post-Event," *PloS Currents* 7 (October 26, 2015), https://www .ncbi.nlm.nih.gov/pmc/articles/PMC4639320/.

42. Melinda Smith, MA; Lawrence Robinson; and Jeanne Segal, PhD, "Helping Children Cope with Traumatic Events," HelpGuide.org, April 2021, https://www.helpguide .org/articles/ptsd-trauma/helping-children-cope-with-traumatic-stress.htm.

43. "Understanding the Impact of Trauma," chapter 3 in *Trauma-Informed Care in Behavioral Health Services* (Rockville, MD: Substance Abuse and Mental Health Services Administration, 2014), https://www.ncbi.nlm.nih.gov/books/NBK207191/.

44. Susanta Kumar Padhy et al., "Mental Health Effects of Climate Change," *Indian Journal of Occupational & Environmental Medicine* 19, no. 1 (January–April 2015): 3–7, https:// www.ncbi.nlm.nih.gov/pmc/articles/PMC4446935/.

45. Baher Kamal, "Climate Migrants Might Reach One Billion by 2050," ReliefWeb, August 21, 2017, https://reliefweb.int/report/world/climate-migrants-might-reach -one-billion-2050.

46. Eleanor Goldberg, "Climate Change Could Have a Serious Impact On Mental Health: Report," *Huffpost*, April 3, 2017, https://www.huffpost.com/entry/climate -change-mental-health_n_58dd584fe4b0e6ac70936be2.

47. Kim Parker et al., "7. Life Satisfaction and Social Support in Different Communities," Pew Research Center, May 22, 2018, https://www.pewsocialtrends.org/2018/05/22 /life-satisfaction-and-social-support-in-different-communities/.

48. Whitmore Williams et al., "Mental Health and Our Changing Climate."

49. Ossie Michelin, "'Solastalgia': Arctic Inhabitants Overwhelmed by New Form of Climate Grief," *Guardian*, October 15, 2020, https://www.theguardian.com/us-news /2020/oct/15/arctic-solastalgia-climate-crisis-inuit-indigenous.

50. Rebekah Levine Coley, PhD, ed., "Understanding the Impacts of Natural Disasters on Children," Society for Research in Child Development, August 13, 2020, https:// www.srcd.org/research/understanding-impacts-natural-disasters-children.

51. A. Crimmins et al., eds., "Mental Health and Well-Being," chapter 8 in *The Impacts of Climate Change on Human Health in the United States: A Scientific Assessment* (Washington, DC: US Global Change Research Program, 2016), https://health2016.globalchange .gov/mental-health-and-well-being.

52. "Behavioral Health Conditions in Children."

53. "Coping with Disaster," Ready.gov, December 17, 2020, https://www.ready.gov /coping-disaster.

54. "Behavioral Health Conditions in Children."

55. "One Year after Katrina, More Is Known About Its Mental Health Effects; Storm's Widespread Effect on People of Color and Children and the Need for Culturally Competent Mental Health Services Are Evident," American Psychological Association, 2006, https://www.apa.org/news/press/releases/2006/08/katrina.

56. Suniya S. Luthar, "The Construct of Resilience: A Critical Evaluation and Guidelines for Future Work," *Child Development* 71, no. 3 (2000): 543–62, https://www.ncbi.nlm .nih.gov/pmc/articles/PMC1885202/.

57. Nikunj Makwana, "Disaster and Its Impact on Mental Health: A Narrative Review," *Journal of Family Medicine and Primary Care* 8, no. 10 (October 2019): 3090–95, https:// www.ncbi.nlm.nih.gov/pmc/articles/PMC6857396/.

58. "Building Your Resilience," American Psychological Association, 2012, https://www .apa.org/topics/resilience.

59. Scott Berson, "What Are Zello and Firechat, the Apps People Are Downloading for Hurricane Florence?" *Charlotte Observer*, September 12, 2018, https://www .charlotteobserver.com/latest-news/article218248820.html.

60. Holly Hartman, "I Downloaded an App and Suddenly Was Part of the Cajun Navy," *Houston Chronicle*, September 4, 2017, https://www.houstonchronicle.com/local/gray-matters/article/I-downloaded-an-app-And-suddenly-I-was-talking-12172506.php.

61. Brandi Neal, "These Apps Can Save Your Life During a Disaster," Bustle.com, September 11, 2017, https://www.bustle.com/p/emergency-apps-you-can-use-without-wi-fi-during-a-disaster-2307076.

62. Berson, "What Are Zello and Firechat."

63. "PawBoost—Lost and Found Pets," App Store Preview, Apple.com, https://apps.apple.com/us/app/pawboost-lost-and-found-pets/id1131343030.

64. Katie Thornton, "Building Resilience in the Aftermath of Natural Disasters," Karyn Purvis Institute of Child Development, TCU College of Science & Engineering, https://child.tcu.edu/building-resilience-in-the-aftermath-of-natural-disasters/.

65. "Gardens at the Stowe Center," Harriet Beecher Stowe Center, https://www.harrietbeecherstowecenter.org/visit/gardens/.

66. TKF Foundation, "Short Documentary Film Tells a Different Kind of Hurricane Sandy Story," Cision, October 27, 2017, https://www.prnewswire.com/news-releases/short-documentary-film-tells-a-different-kind-of-hurricane-sandy-story-300544725.html.

67. "Hurricane Healing Garden," Nature Sacred, https://naturesacred.org/films/hurricane-healing-garden/.

68. Ibid.

69. Keith G. Tidball, "Urgent Biophilia: Human–Nature Interactions and Biological Attractions in Disaster Resilience," *Ecology and Society* 17, no. 2 (2012), https://www.ecologyandsociety.org/vol17/iss2/art5/.

70. Kara Rogers, "Biophilia hypothesis," *Encyclopedia Britannica*, June 25, 2019, https://www.britannica.com/science/biophilia-hypothesis.

71. TKF Foundation, "Short Documentary Film."

72. "Aftershocks of Disaster: Puerto Rico Before and After the Storm," Haymarket Books, August 27, 2020, https://www.haymarketbooks.org/events/178-aftershocks-of-disaster-puerto-rico-before-and-after-the-storm.

73. Ibid.

74. Ibid.

75. Tara Haelle, MS, "Mindfulness-Based Stress Reduction Therapy Holds Promise for PTSD Treatment," Psychiatry Advisor, December 5, 2017, https://www.psychiatryadvisor.com/home/topics/anxiety/ptsd-trauma-and-stressor-related/mindfulness-based-stress-reduction-therapy-holds-promise-for-ptsd-treatment/.

76. James Douglas Bremner et al., "A Pilot Study of the Effects of Mindfulness-Based Stress Reduction on Post-Traumatic Stress Disorder Symptoms and Brain Response to Traumatic Reminders of Combat in Operation Enduring Freedom/Operation Iraqi Freedom Combat Veterans with Post-Traumatic Stress Disorder," *Frontiers in Psychiatry* 8 (2017), https://www.frontiersin.org/articles/10.3389/fpsyt.2017.00157/full.

77. Jenna E. Boyd, MSc; Ruth A. Lanius, MD, PhD; and Margaret C. McKinnon, PhD, CPsych, "Mindfulness-Based Treatments for Posttraumatic Stress Disorder: A Review of the Treatment Literature and Neurobiological Evidence," *Journal of Psychiatry & Neuroscience* 43, no. 1 (January 2018): 7–25, https://www.ncbi.nlm.nih.gov/pmc/articles/PMC5747539/.

78. David J. Kearney et al., "Loving-Kindness Meditation for Posttraumatic Stress Disorder: A Pilot Study," *Journal of Traumatic Stress* 26, no. 4 (August 2013): 426–34, https://pubmed.ncbi.nlm.nih.gov/23893519/.

79. "About Jack Kornfield," JackKornfield.com, https://jackkornfield.com/bio/.

80. "A Meditation on Lovingkindness," JackKornfield. com, https://jackkornfield.com/meditation-lovingkindness/.

81. Carolyn L. Todd, "The Healing Powers of Gratitude," Self.com, May 29, 2020, https://www.self.com/story/gratitude-benefits.

82. Wenchao Wang, Xinchun Wu, and Yuxin Tian, "Mediating Roles of Gratitude and Social Support in the Relation between Survivor Guilt and Posttraumatic Stress Disorder, Posttraumatic Growth among Adolescents after the Ya'an Earthquake," *Frontiers in Psychology* 9 (November 5, 2018), https://www.ncbi.nlm.nih.gov/pmc/articles/PMC6230928/.

83. Todd, "The Healing Powers of Gratitude."

84. "Giving Thanks Can Make You Happier," Harvard Health Publishing, November 22, 2011, https://www.health.harvard.edu/healthbeat/giving-thanks-can -make-you-happier.

85. Madhuleena Roy Chowdhury, BA, "The Neuroscience of Gratitude and How It Affects Anxiety & Grief," PositivePsychology.com, May 26, 2021, https://positivepsy chology.com/neuroscience-of-gratitude/.

86. "4 Tips for Keeping a Gratitude Journal," Cleveland Clinic, July 20, 2020, https:// health.clevelandclinic.org/tips-for-keeping-a-gratitude-journal/.

87. "Loving-Kindness Meditation," Greater Good in Action, https://ggia.berkeley.edu /practice/loving_kindness_meditation.

Chapter Three
Cities of Heat

1. Associated Press, "France Says 1,500 Died in Summer's Heat Wave," ABC News, September 8, 2019, https://abcnews.go.com/Health/wireStory/france-1500-died -summers-heat-wave-65466010.

2. "Extreme Heat," Ready.gov, May 26, 2021, https://www.ready.gov/heat.

3. Alissa Walker, "Our Cities Are Getting Hotter—and It's Killing People," Curbed, June 21, 2019, https://archive.curbed.com/2018/7/6/17539904/heat-wave-extreme -heat-cities-deadly.

4. "NC Climate Education," North Carolina State University, https://climate.ncsu.edu /edu/Albedo.

5. "American Warming: The Fastest-Warming Cities and States in the US," Climate Central, April 17, 2019, https://www.climatecentral.org/news/report-american -warming-us-heats-up-earth-day.

6. "Street, Inlets, and Storm Drains," chapter 7 in *Urban Storm Drainage Criteria Manual*, vol. 1 (Denver: Urban Drainage and Flood Control District, 2016), https://udfcd.org /wp-content/uploads/uploads/vol1%20criteria%20manual/07_Streets%20Inlets%20 Storm%20Drains.pdf.

7. Allegra Miccio, "Wastewater Fees Encourage Green Infrastructure Initiatives in NYC," *Water Watch NYC* (blog), January 9, 2018, https://waterwatchnyc.com/tag/impervious-surfaces/.

8. "Dallas Urban Heat Island Mitigation Study," Texas Trees Foundation, https://www.texastrees.org/projects/dallas-urban-heat-island-mitigation-study/.

9. "Nights Warming Faster Than Days across Much of the Planet," University of Exeter, October 1, 2020, https://www.exeter.ac.uk/news/homepage/title_818402_en.html.

10. Peninah Murage, Shakoor Hajat, and R. Sari Kovats, "Effect of Night-Time Temperatures on Cause and Age-Specific Mortality in London," *Environmental Epidemiology* 1, no. 2 (December 2017): e005, https://journals.lww.com/environepidem/fulltext/2017/12000/effect_of_night_time_temperatures_on_cause_and.1.aspx.

11. "Ground-Level Ozone Basics," Environmental Protection Agency, https://www.epa.gov/ground-level-ozone-pollution/ground-level-ozone-basics.

12. Ibid.

13. Kim Parker et al., "Demographic and Economic Trends in Urban, Suburban, and Rural Communities," Pew Research Center, May 22, 2018, https://www.pewresearch.org/social-trends/2018/05/22/demographic-and-economic-trends-in-urban-suburban-and-rural-communities/.

14. Richard Florida, "Where Kids Live Now in the US," Bloomberg, April 13, 2015, https://www.bloomberg.com/news/articles/2015-04-13/the-u-s-metros-with-the-highest-and-lowest-shares-of-children-under-18.

15. "Never Leave Your Child Alone in a Car: Frequently Asked Questions," Safe Kids Worldwide, https://www.safekids.org/sites/default/files/documents/activity_kits/heat stroke/heatstroke_faqs_2015.pdf.

16. Protecting Your Child from Dehydration and Heat Illness," WebMD, October 5, 2019, https://www.webmd.com/children/dehydration-heat-illness.

17. "How Much Hotter Is Your Hometown Than When You Were Born?" *New York Times*, August 30, 2018, https://www.nytimes.com/interactive/2018/08/30/climate/how-much-hotter-is-your-hometown.html.

18. Ibid.

19. "Heat," *Health Hints* 10, no. 8 (September 2006), https://texashelp.tamu.edu/wp
-content/uploads/2016/02/heat-health-hints.pdf.

20. "2019 Profile of Older Americans," Administration for Community Living, US
Department of Health and Human Services, May 2020, https://acl.gov/sites/default
/files/Aging%20and%20Disability%20in%20America/2019ProfileOlderAmeric
ans508.pdf.

21. Joel Kotkin, "America's Senior Moment: The Most Rapidly Aging Cities," *Forbes*,
February 16, 2016, https://www.forbes.com/sites/joelkotkin/2016/02/16/americas
-senior-moment-the-most-rapidly-aging-cities/?sh=46d8dfec53e5.

22. Jacob Ausubel, "Older People Are More Likely to Live Alone in the US Than Else-
where in the World," Pew Research Center, March 10, 2020, https://www.pewre
search.org/fact-tank/2020/03/10/older-people-are-more-likely-to-live-alone-in
-the-u-s-than-elsewhere-in-the-world/.

23. Ibid.

24. Ibid.

25. Laura Silver et al., "In US and UK, Globalization Leaves Some Feeling 'Left Behind'
or 'Swept Up,'" Pew Research Center, October 5, 2020, https://www.pewresearch
.org/2020/10/05/in-u-s-and-uk-globalization-leaves-some-feeling-left-behind-or
-swept-up/.

26. Craille Maguire Gillies, "What's the World's Loneliest City?" *Guardian*, April 7, 2016,
https://www.theguardian.com/cities/2016/apr/07/loneliest-city-in-world.

27. Ibid.

28. "Social Isolation, Loneliness in Older People Pose Health Risks," National Institute
on Aging, April 23, 2019, https://www.nia.nih.gov/news/social-isolation-loneliness
-older-people-pose-health-risks.

29. "Fighting Social Isolation among Older Adults," GreatCall, https://www.greatcall
.com/docs/default-source/newsroom-files/fighting-social-isolation-among-older
-adults.pdf.

30. Marco Chown Oved, "Life and Death under the Dome," *Toronto Star*, May 23, 2019, https://projects.thestar.com/climate-change-canada/quebec.

31. Charlie Fidelman, "As Second Heat Wave Gains Steam, 74 Deaths Are Linked to Quebec Weather," *Montreal Gazette*, July 10, 2018, https://montrealgazette.com /news/quebec/montreal-heat-wave-more-searing-weather-in-the-forecast-next -weekend.

32. Brittany Henriques, "66 Montrealers Died from Extreme Heat during 2018 Heat Wave," Global News, August 12, 2019, https://globalnews.ca/news/5279502/66 -montrealers-died-from-extreme-heat-during-2018-heat-wave/.

33. Chown Oved, "Life and Death under the Dome."

34. Elzbieta Sawicz, "Summertime Control of Temperature in Canadian Homes: How Canadians Keep Their Cool," Statistics Canada, November 27, 2015, https://www150 .statcan.gc.ca/n1/pub/16-002-x/2011002/part-partie3-eng.htm.

35. Malcolm Araos, "Climate Change Can Be Deadly If You Live Alone," *Conversation*, August 1, 2018, https://theconversation.com/climate-change-can-be-deadly-if-you -live-alone-100881.

36. Malcolm Araos, "Men Who Live Alone Are Most Likely to Die during Heat Waves," City Monitor, August 6, 2018, https://citymonitor.ai/horizons/men-who-live-alone -are-most-likely-die-during-heatwaves-4110.

37. "NOAA Leads Community Scientists in Mapping Hottest Parts of 13 US Cities This Summer," Climate Program Office, June 3, 2020, https://cpo.noaa.gov/News /ArtMID/7875/ArticleID/1922/NOAA-leads-community-scientists-in-mapping -hottest-parts-of-13-US-cities-this-summer.

38. G. Brooke Anderson and Michelle L. Bell, "Heat Waves in the United States: Mortality Risk during Heat Waves and Effect Modification by Heat Wave Characteristics in 43 US Communities," Michelle L. Bell's Research Group, Yale University, 2010, https:// bell-lab.yale.edu/publications/heat-waves-united-states-mortality-risk-during-heat -waves-and-effect-modification-heat.

39. Tim De Chant, "US Cities Are Losing 36 Million Trees a Year," *Nova*, May 8, 2018, https://www.pbs.org/wgbh/nova/article/us-urban-forest-crisis/.

40. "For Tree Equity and Climate Change, How Many Urban Trees Do We Need?" American Forests, https://www.americanforests.org/our-work/urban-forestry/how -many-urban-trees-do-we-need/.

41. Josh Wood, "Re-Greening: Can Louisville Plant Its Way Out of a Heat Emergency?" *Guardian*, November 21, 2019, https://www.theguardian.com/cities/2019/nov/21 /re-greening-can-louisville-plant-its-way-out-of-a-heat-emergency.

42. Darcy Costello, "Tree Canopy Ordinance Finally Clears Metro Council, Along with Rental Help, City Borrowing," *Louisville Courier Journal*, April 24, 2020, https://www .courier-journal.com/story/news/politics/metro-government/2020/04/24/tree -canopy-louisville-metro-council-oks-legislation-curb-losses/3018922001/.

43. "The Benefits of Trees," Canopy, https://canopy.org/tree-info/benefits-of-trees/.

44. Meg Anderson and Sean McMinn, "As Rising Heat Bakes US Cities, the Poor Often Feel It Most," NPR, September 3, 2019, https://www.npr.org/2019/09/03/75404 4732/as-rising-heat-bakes-u-s-cities-the-poor-often-feel-it-most.

45. Ibid.

46. "2021 Demographics," Miami Matters, http://www.miamidadematters.org/demo graphicdata.

47. "The Racial Wealth Divide in Miami," Prosperity Now, https://prosperitynow.org /files/resources/Racial_Wealth_Divide_in_Miami_OptimizedforScreenReaders.pdf.

48. "The Heat Is On: US Temperature Trends," Climate Central, June 23, 2012, https:// www.climatecentral.org/news/the-heat-is-on/.

49. "Little Haiti Community Needs Assessment," Children's Trust, 2015, https://www .thechildrenstrust.org/sites/default/files/kcfinder/files/providers/analytics/reports /Little_Haiti_Miami-Dade_County-May2015.pdf.

50. "Little Haiti Community Needs Assessment: Housing Market Analysis," Haitian American Community Development Corporation, December 2015, https:// metropolitan.fiu.edu/research/services/economic-and-housing-market-analysis /2015-hacdc_finalreport.pdf.

51. Allison Rebecca Penn, "Landlords, Are You Responsible for Air-Conditioning?" All Property Management, June 17, 2019, https://www.allpropertymanagement.com /blog/post/landlord-responsiblities-for-air-conditioning/.

52. "Climate Change and Displacement in the US—A Review of the Literature," Urban Displacement Project, April 2020, https://www.urbandisplacement.org/sites/default /files/images/climate_and_displacement_-_lit_review_6.19.2020.pdf.

53. "Household Expenditures and Income," Pew Charitable Trusts, March 2016, https:// www.pewtrusts.org/-/media/assets/2016/03/household_expenditures_and_income.pdf.

54. "Money for Heat and Utility Expenses," Access NYC, https://access.nyc.gov /programs/home-energy-assistance-program-heap/.

55. María Paula Rubiano, "In Los Angeles, Rich Neighborhoods Enjoy More Street Trees and a Lot More Birds," Audubon, July 7, 2020, https://www.audubon.org/news/in -los-angeles-rich-neighborhoods-enjoy-more-street-trees-and-lot-more-birds.

56. Tim De Chant, "Urban Trees Reveal Income Inequality," *Per Square Mile* (blog), May 17, 2012, https://persquaremile.com/2012/05/17/urban-trees-reveal-income -inequality/.

57. "The Benefits of Trees."

58. Matt O'Brien, "Gauge of Bay Area Neighborhood Wealth? Look to the Trees," *Mercury News*, September 9, 2012, https://www.mercurynews.com/2012/09/09/gauge-of -bay-area-neighborhood-wealth-look-to-the-trees/.

59. Kirsten Schwarz et al., "Trees Grow on Money: Urban Tree Canopy Cover and Environmental Justice," *PLoS ONE* 10, no. 4 (April 1, 2015), https://journals.plos .org/plosone/article?id=10.1371/journal.pone.0122051.

60. Ibid.

61. Ibid.

62. Patrick Sisson, Jeff Andrews, and Alex Bazeley, "The Affordable Housing Crisis, Explained," Curbed, March 2, 2020, https://archive.curbed.com/2019/5/15/18617763 /affordable-housing-policy-rent-real-estate-apartment.

63. "Report: Coastal Flood Risk to Affordable Housing Projected to Triple by 2050," Climate Central, November 24, 2020, https://www.climatecentral.org/news/report -coastal-flood-risk-to-affordable-housing-projected-to-triple-by-2050.

64. Patrick Sisson, "In Many Cities, Climate Change Will Flood Affordable Housing," CityLab, December 1, 2020, https://www.bloomberg.com/news/articles /2020-12-01/how-climate-change-is-targeting-affordable-housing.

65. Portland State University, "Historical Housing Disparities Linked with Dangerous Climate Impacts," ScienceDaily, January 14, 2020, https://www.sciencedaily.com /releases/2020/01/200114101715.htm.

66. Gwen Sharp, PhD, "Philadelphia Redlining Maps," Society Pages, April 25, 2012, https://thesocietypages.org/socimages/2012/04/25/1934-philadelphia-redlining -map/.

67. Ibid.

68. Tracy Jan, "Redlining Was Banned 50 Years Ago. It's Still Hurting Minorities Today," *Washington Post*, March 28, 2018, https://www.washingtonpost.com/news/wonk /wp/2018/03/28/redlining-was-banned-50-years-ago-its-still-hurting-minorities -today/.

69. Jenny Rowland-Shea et al., "The Nature Gap," Center for American Progress, July 21, 2020, https://www.americanprogress.org/issues/green/reports/2020/07/21/487787 /the-nature-gap/.

70. "Redlining—and Greening—of Cities. What's the Connection?" *Tree Equity* (blog), January 16, 2020, https://www.americanforests.org/blog/redlining-and-greening-of -cities-whats-the-connection/.

71. "Benefits of Urban Trees," South Carolina Forestry Commission, https://www.state .sc.us/forest/urbben.htm.

72. "Trees in Urban Areas May Improve Mental Health," European Commission, April 16, 2015, https://ec.europa.eu/environment/integration/research/newsalert /pdf/trees_in_urban_areas_may_improve_mental_health_410na2_en.pdf.

73. "Tree Equity Score," American Forests, https://www.treeequityscore.org/.

74. Eriol Iasa, "How to Bridge the 'Canopy Gap'? Counting and Mapping Trees," blog post, November 20, 2020, https://eriolasa.blogspot.com/2020/11/how-to-bridge-canopy-gap-counting-and.html.

75. "Tree Equity in America's Cities," American Forests, https://www.americanforests.org/our-work/urban-forestry/.

76. "US Chapter Launch," 1t.org, https://us.1t.org/press-toolkit/.

77. "Secretary Chu Announces Steps to Implement Cool Roofs at DOE and Across the Federal Government," US Department of Energy, July 19, 2010, https://www.energy.gov/articles/secretary-chu-announces-steps-implement-cool-roofs-doe-and-across-federal-government.

78. "Cool Roofs," Berkeley Lab, https://heatisland.lbl.gov/coolscience/cool-roofs.

79. Ibid.

80. "Dallas Urban Heat Management Study," Texas Trees Foundation, 2017, https://urbanforestrysouth.org/resources/library/ttresources/dallas-urban-heat-management-study/at_download/file.

81. "Cool Pavement Pilot Program," City of Phoenix, https://www.phoenix.gov/streets/coolpavement.

82. "Case Study: Deadly Chicago Heat Wave of 1995, AdaptNY, https://www.adaptny.org/2016/07/21/case-study-deadly-chicago-heat-wave-of-1995/.

83. Doug Garrett, "Beware the Closed Bedroom Door," *Home Energy*, January 1, 2001, http://homeenergy.org/show/article/id/1704.

84. Samantha Leffler, "7 Ways to Keep Cool in NYC Without Air-Conditioning," Brick Underground, July 19, 2017, https://www.brickunderground.com/live/how-to-cool-down-without-Air-conditioning-in-NYC.

Chapter Four:
Viruses and Infectious Diseases—from Corona to Lyme

1. "WHO Timeline—COVID-19," World Health Organization, April 27, 2020, https://www.who.int/news/item/29-06-2020-covidtimeline.

2. David M. Morens and Anthony S. Fauci, "Emerging Pandemic Diseases: How We Got to COVID-19," *Cell* 182, no. 5 (September 3, 2020): 1077–92, https://pubmed.ncbi.nlm.nih.gov/32846157/.

3. Ibid.

4. John S. Mackenzie and David W. Smith, "COVID-19: A Novel Zoonotic Disease Caused by a Coronavirus from China: What We Know and What We Don't," *Microbiology Australia* (March 17, 2020): MA20013, https://www.ncbi.nlm.nih.gov/pmc/articles/PMC7086482/.

5. Ben Westcott and Serenitie Wang, "China's Wet Markets Are Not What Some People Think They Are," CNN, April 23, 2020, https://www.cnn.com/2020/04/14/asia/china-wet-market-coronavirus-intl-hnk/index.html.

6. Sumi Krishna, "Don't Fear or Judge the Wet Market Too Quickly," Wire: Science, June 13, 2020, https://science.thewire.in/environment/coronavirus-pandemic-wet-markets-wild-foods/.

7. Robert G. Webster, "Wet Markets—A Continuing Source of Severe Acute Respiratory Syndrome and Influenza?" *Lancet* 363, no. 9404 (January 17, 2004): 234–36, https://pubmed.ncbi.nlm.nih.gov/14738798/.

8. Rui-Heng Xu et al., "Epidemiologic Clues to SARS Origin in China," *Emerging Infectious Diseases* 10, no. 6 (June 2004): 1030–37, https://www.ncbi.nlm.nih.gov/pmc/articles/PMC3323155/.

9. Jonathan Watts, "China Culls Wild Animals to Prevent New SARS Threat," *Lancet* 363 (January 10, 2004): 134, https://www.thelancet.com/pdfs/journals/lancet/PIIS0140-6736(03)15313-5.pdf.

10. "Dr. Fauci on Change in Face Mask Recommendations, Using Hydroxychloroquine as a COVID-19 Treatment," Fox News, April 3, 2020, https://video.foxnews.com/v/6146838481001/.

11. Alla Katsnelson, "How Do Viruses Leap from Animals to People and Spark Pandemics?" *Chemical & Engineering News* 98, no. 33 (August 30, 2020), https://cen.acs.org /biological-chemistry/infectious-disease/How-do-viruses-leap-from-animals-to -people-and-spark-pandemics/98/i33.

12. Richard Tapper, "Wildlife Watching and Tourism: A Study on the Benefits and Risks of a Fast-Growing Tourism Activity and Its Impacts on Species," United Nations Environment Programme, January 2006, https://www.researchgate.net /publication/268036486_Wildlife_Watching_and_Tourism_A_Study_on_the _Benefits_and_Risks_of_a_Fast_Growing_Tourism_Activity_and_its_Impacts_on _Species.

13. Chris Mooney, "Natural Protected Areas Get 8 Billion Visits per Year. That's Higher Than the World's Population," *Washington Post*, February 24, 2015, https://www .washingtonpost.com/news/energy-environment/wp/2015/02/24/new-study -proves-that-human-beings-love-nature/.

14. Robert W. Sutherst, "Global Change and Human Vulnerability to Vector-Borne Diseases," *Clinical Microbiology Reviews* 17, no. 1 (January 2004): 136–73, https://www .ncbi.nlm.nih.gov/pmc/articles/PMC321469/.

15. Helen Briggs, "Covid: Why Bats Are Not to Blame, Say Scientists," BBC News, October 13, 2020, https://www.bbc.com/news/science-environment-54246473.

16. "Ebola Virus Disease," World Health Organization, February 23, 2021, https://www .who.int/news-room/fact-sheets/detail/ebola-virus-disease.

17. "Guinea Declares Ebola Epidemic: First Deaths Since 2016," BBC News, February 14, 2021, https://www.bbc.com/news/world-africa-56060728?fbclid=IwAR2mj LMvxh5J8IZzObCu4n2Kt5fT9dZLO0w3TTHpXMOgOnxBqWKUtpxEBKU.

18. Robert Nasi, "Conservation and Use of Wildlife-Based Resources: The Bushmeat Crisis," Secretariat of the Convention on Biological Diversity, 2008, https://www .academia.edu/3261342/Conservation_and_use_of_wildlife_based_resources_the _bushmeat_crisis.

19. Michael Dulaney, "The Next Pandemic Is Coming—and Sooner Than We Think, Thanks to Changes to the Environment," ABC News (Australia), June 6, 2020, https://www.abc.net.au/news/science/2020-06-07/a-matter-of-when-not-if-the -next-pandemic-is-around-the-corner/12313372.

20. "Cardiovascular Epidemiologist Christopher Reid," ABC News (Australia), April 5, 2020, https://www.abc.net.au/news/2020-04-06/cardiovascular-epidemiologist-and-clinical-trialist-christopher/12124710.

21. "Deforestation and Forest Degradation," World Wildlife Fund, https://www.worldwildlife.org/threats/deforestation-and-forest-degradation.

22. Ibid.

23. "NASA Tropical Deforestation Research," NASA Earth Observatory," March 30, 2007, https://earthobservatory.nasa.gov/features/Deforestation/deforestation_update4.php.

24. "Where Is the World's Biological Diversity Found?" Sinauer Associates Inc., 2010, https://www.sinauer.com/media/wysiwyg/samples/PrimackEssentials5e_Ch03.pdf.

25. Robert Kessler, "What Exactly Is Deforestation Doing to Our Planet?" EcoHealth Alliance, https://www.ecohealthalliance.org/2017/11/deforestation-impact-planet.

26. Jesús Olivero et al., "Recent Loss of Closed Forests Is Associated with Ebola Virus Disease Outbreaks," *Scientific Reports* 7 (2017): 14291, https://www.ncbi.nlm.nih.gov/pmc/articles/PMC5662765/.

27. Nellie Peyton, "Want to Know When Ebola Will Strike Next? Look to the Forest," Reuters, October 30, 2017, https://www.reuters.com/article/us-africa-disease-ebola/want-to-know-when-ebola-will-strike-next-look-to-the-forest-idUSKBN1CZ236.

28. Felicia Keesing et al., "Impacts of Biodiversity on the Emergence and Transmission of Infectious Diseases," *Nature* 468 (2010): 647–52, https://www.nature.com/articles/nature09575.

29. Matthew B. Thomas, "Epidemics on the Move: Climate Change and Infectious Disease," *PloS Biology* 18, no. 11 (November 2020): e3001013, https://www.ncbi.nlm.nih.gov/pmc/articles/PMC7685491/.

30. "Media Release: Nature's Dangerous Decline 'Unprecedented'; Species Extinction Rates 'Accelerating,'" Intergovernmental Science-Policy Platform on Biodiversity and Ecosystem Services, https://ipbes.net/news/Media-Release-Global-Assessment.

31. "Halting the Extinction Crisis," Center for Biological Diversity, https://www.biologicaldiversity.org/programs/biodiversity/elements_of_biodiversity/extinction_crisis/.

32. "Illnesses from Mosquito, Tick, and Flea Bites Increasing in the US," Centers for Disease Control and Prevention, May 1, 2018, https://www.cdc.gov/media/releases/2018/p0501-vs-vector-borne.html.

33. Mohd Danish Khan et al., "Aggravation of Human Diseases and Climate Change Nexus," *International Journal of Environmental Research and Public Health* 16, no. 15 (August 2019): 2799, https://www.ncbi.nlm.nih.gov/pmc/articles/PMC6696070/.

34. Julia Langer, Abbey Dufoe, and Jen Brady, "US Faces a Rise in Mosquito 'Disease Danger Days,'" Climate Central, August 8, 2018, http://assets.climatecentral.org/pdfs/August2018_CMN_Mosquitoes.pdf?pdf=Mosquitoes-Report.

35. Ibid.

36. "Lyme Disease Diagnostics Research," National Institute of Allergy and Infectious Diseases, November 20, 2018, https://www.niaid.nih.gov/diseases-conditions/lyme-disease-diagnostics-research.

37. "Justin Wood, Founder of Geneticks," Looking at Lyme, July 28, 2020, https://www.lookingatlyme.ca/2020/07/justin-wood-founder-of-geneticks/.

38. Alison Karlene Hodgins, "Bitten by a Tick in Canada? Here's How to Get It Tested for Lyme Disease—Fast," *Explore*, August 19, 2020, https://www.explore-mag.com/Bitten-by-a-tick-in-Canada-Heres-how-to-get-it-tested-for-Lyme-disease-fast.

39. Catherine Caruso, "Tests for Lyme Disease Miss Many Early Cases—but a New Approach Could Help," Stat, June 28, 2017, https://www.statnews.com/2017/06/28/early-lyme-tests/.

40. Hodgins, "Bitten by a Tick in Canada."

41. Ibid.

42. "About Us," Geneticks, https://www.geneticks.ca/about-geneticks/.

43. Ibid.

44. Hodgins, "Bitten by a Tick in Canada."

45. "Rocky Mountain Spotted Fever," National Institute of Allergy and Infectious Diseases, July 8, 2014, https://www.niaid.nih.gov/diseases-conditions/rocky-mountain-spotted-fever.

46. Timothy C. Winegard, *The Mosquito: A Human History of Our Deadliest Predator* (New York: Dutton, 2020).

47. Ibid.

48. "West Nile Virus," Centers for Disease Control and Prevention, https://www.cdc .gov/westnile/index.html.

49. Ibid.

50. Andrea Thompson, "What Warming Means for 4 of Summer's Worst Pests," Climate Central, July 30, 2015, https://www.climatecentral.org/news/what-warming-means -summers-pests-19295.

51. "Chikungunya," World Health Organization, September 15, 2020, https://www.who .int/news-room/fact-sheets/detail/chikungunya.

52. Shlomit Paz, "Climate Change Impacts on West Nile Virus Transmission in a Global Context," *Philosophical Transactions B* 370, no. 1665 (April 5, 2015): 20130561, https:// www.ncbi.nlm.nih.gov/pmc/articles/PMC4342965/.

53. "Asian Tiger Mosquito," National Invasive Species Information Center, https://www .invasivespeciesinfo.gov/terrestrial/invertebrates/asian-tiger-mosquito.

54. Philip M. Armstrong et al., "Northern Range Expansion of the Asian Tiger Mosquito (*Aedes albopictus*): Analysis of Mosquito Data from Connecticut, USA," *PloS Neglected Tropical Diseases* 11, no. 5 (May 18, 2017): e0005623, https://journals.plos.org /plosntds/article?id=10.1371/journal.pntd.0005623.

55. Carl Zimmer, "West Nile Virus: The Stranger That Came to Stay," *Discover*, August 17, 2012, https://www.discovermagazine.com/planet-earth/west-nile-virus-the-stranger -that-came-to-stay.

56. "West Nile Virus," One Health Commission, November 21, 2011, https://www .onehealthcommission.org/index.cfm/38649/15886/west_nile_virus.

57. "The History of Zika Virus," World Health Organization, February 7, 2016, https:// www.who.int/news-room/feature-stories/detail/the-history-of-zika-virus.

58. "Fifth Meeting of the Emergency Committee under the International Health Regulations (2005) Regarding Microcephaly, Other Neurological Disorders, and Zika Virus," Pan

American Health Organization, November 18, 2016, https://www3.paho.org/hq/index.php?option=com_content&view=article&id=12761:v-meetingemergency-committee-ihr-2005-microcephaly-neurological-dis-zika&Itemid=135&lang=en.

59. Ann Nagro, "State of the Mosquito Market Report," MGK, 2017, https://www.mgk.com/wp-content/uploads/2019/09/2017-PCT-State-of-the-Mosquito-Market-Report.pdf.

60. Sarah Anderson, "The Mosquito Gap," OtherWords, March 28, 2018, https://otherwords.org/the-mosquito-gap/.

61. John P. Roche, "Study Finds Bigger Mosquitoes in Baltimore Neighborhoods with More Abandoned Buildings," *Entomology Today*, October 24, 2019, https://entomologytoday.org/2019/10/24/study-finds-bigger-mosquitoes-in-baltimore-neighborhoods-with-more-abandoned-buildings/.

62. Shannon L. LaDeau et al., "Higher Mosquito Production in Low-Income Neighborhoods of Baltimore and Washington, DC: Understanding Ecological Drivers and Mosquito-Borne Disease Risk in Temperate Cities," *International Journal of Environmental Research and Public Health* 10, no. 4 (April 2013): 1505–26, https://www.ncbi.nlm.nih.gov/pmc/articles/PMC3709331/.

63. Sandra Crouse Quinn and Supriya Kumar, "Health Inequalities and Infectious Disease Epidemics: A Challenge for Global Health Security," *Biosecurity and Bioterrorism: Biodefense Strategy, Practice, and Science* 12, no. 5 (September 1, 2014): 263–73, https://www.ncbi.nlm.nih.gov/pmc/articles/PMC4170985/.

64. Thomas M. Kollars, "Identifying High-Risk Areas of West Nile Virus in Minority and Low-Income Neighborhoods," *Clinical Microbiology and Infectious Diseases* 2, no. 1 (2017): 1–3, https://www.oatext.com/Identifying-high-risk-areas-of-West-Nile-Virus-in-minority-and-low-income-neighborhoods.php.

65. Donald Thea, "How to Contain Zika in the US," Boston University School of Public Health, April 29, 2016, https://www.bu.edu/sph/news/articles/2016/viewpoint-how-to-contain-zika-in-the-us/.

66. "Prevent Mosquito Bites," Centers for Disease Control and Prevention, https://www.cdc.gov/ncezid/dvbd/media/stopmosquitoes.html.

67. "Researchers Discover How Mosquitoes Smell Human Sweat," National Institute of Allergy and Infectious Diseases, July 30, 2019, https://www.niaid.nih.gov/news-events/how-mosquitoes-smell-human-sweat.

68. Carolyn S. McBride, "Genes and Odors Underlying the Recent Evolution of Mosquito Preference for Humans," *Current Biology* 26, no. 1 (January 11, 2016): R41–R46, doi: 10.1016/j.cub.2015.11.032.

69. "Researchers Discover How Mosquitoes Smell Human Sweat."

70. Cameron Webb, "Feel Like You're a Mozzie Magnet? It's True—Mosquitoes Prefer to Bite Some People Over Others," *Conversation*, February 10, 2020, https://the conversation.com/feel-like-youre-a-mozzie-magnet-its-true-mosquitoes-prefer-to -bite-some-people-over-others-128788.

71. Cameron Webb, "Health Check: Why Mosquitoes Seem to Bite Some People More," *Conversation*, January 25, 2015, https://theconversation.com/health-check-why -mosquitoes-seem-to-bite-some-people-more-36425.

72. Cheryl Bennett Wiygul, Facebook post, July 10, 2019, https://www.facebook.com /cheryl.bennettwiygul/posts/2604638246213205.

73. Ibid.

74. Ibid.

75. Ibid.

76. "*Vibrio* Species Causing Vibriosis," Centers for Disease Control and Prevention, https://www.cdc.gov/vibrio/index.html.

77. Michael H. Bross, MD, et al., "*Vibrio vulnificus* Infection: Diagnosis and Treatment," *American Family Physician*, August 15, 2007, https://www.aafp.org/afp/2007/0815 /p539.html.

78. Wiygul Facebook post.

79. "People at Risk," Centers for Disease Control and Prevention, https://www.cdc.gov /vibrio/people-at-risk.html.

80. Reem Deeb et al., "Impact of Climate Change on *Vibrio vulnificus* Abundance and Exposure Risk," *Estuaries and Coasts* 41, no. 8 (December 2018): 2289–303, https:// www.ncbi.nlm.nih.gov/pmc/articles/PMC6602088/.

81. Erin K. Lipp, Anwar Huq, and Rita R. Colwell, "Effects of Global Climate on Infectious Disease: The Cholera Model," *Clinical Microbiology Reviews* 15, no. 4 (October 2002): 757–70, https://www.ncbi.nlm.nih.gov/pmc/articles/PMC126864/.

82. "*Vibrio* Background," National Centers for Coastal Ocean Science," https://products .coastalscience.noaa.gov/vibrioforecast/.

83. "Prevention Tips," Centers for Disease Control and Prevention, https://www.cdc .gov/vibrio/prevention.html.

84. "Preventing Tick Bites," Centers for Disease Control and Prevention, https://www .cdc.gov/ticks/avoid/on_people.html.

Chapter Five:
The New Allergy Season

1. "Extreme Allergies and Climate Change," Asthma and Allergy Foundation of America, https://www.aafa.org/extreme-allergies-and-climate-change/.

2. "Climate Change Indicators: Ragweed Pollen Season," Environmental Protection Agency, https://climatechange.stlouis-mo.gov/climate-indicators/climate-change-indicators -ragweed-pollen-season.

3. Ibid.

4. "What Are the Symptoms of an Allergy?" Asthma and Allergy Foundation of America, https://www.aafa.org/allergy-symptoms/.

5. "Allergies and the Immune System," Johns Hopkins Medicine, https://www .hopkinsmedicine.org/health/conditions-and-diseases/allergies-and-the-immune -system#:~:text=Immune%20responses%20can%20be%20mild,of%20the%20 reaction%20may%20increase.

6. "Climate Change Indicators in the United States," Environmental Protection Agency, https://www.epa.gov/climate-indicators.

7. "Spring Coming Earlier (2020)," Climate Central, March 4, 2020, https://medialibrary .climatecentral.org/resources/spring-coming-earlier-2020.

8. Matt Meister, "Spring Leaf-Out Happening Earlier in Colorado," Fox21 News, March 10, 2020, https://www.fox21news.com/weather/thewxmeister-wonders /spring-leaf-out-happening-earlier-in-colorado/.

9. "When to Expect Your Last Spring Freeze," National Centers for Environmental Information, https://www.ncdc.noaa.gov/news/when-expect-your-last-spring-freeze.

10. "Climate Hot Map," Union of Concerned Scientists, 2011, https://www .climatehotmap.org/global-warming-locations/nebraska-usa.html.

11. "Climate Change Indicators in the United States, 2016," Environmental Protection Agency, 2016, https://www.epa.gov/sites/production/files/2016-08/documents /climate_indicators_2016.pdf.

12. "When Asthma Attacks: How to Manage It and Stay Out of the ER," Medical City Healthcare, August 15, 2016, https://medicalcityhealthcare.com/blog/entry/when -asthma-attacks-how-to-manage-it-and-stay-out-of-the-er.

13. James E. Neumann et al., "Estimates of Present and Future Asthma Emergency Department Visits Associated with Exposure to Oak, Birch, and Grass Pollen in the United States," *Geohealth* 3, no. 1 (January 2019): 11–27, https://www.ncbi.nlm.nih .gov/pmc/articles/PMC6516486/.

14. Samantha Harrington, "Climate Change Could Make Children's Allergies Worse," Yale Climate Connections, June 12, 2019, https://yaleclimateconnections.org/2019 /06/climate-change-could-make-childrens-allergies-worse/.

15. "Basic Facts about Mold and Dampness," Centers for Disease Control and Prevention, https://www.cdc.gov/mold/faqs.htm.

16. "Mold Allergy," Asthma and Allergy Foundation of America, https://www.aafa.org /mold-allergy/.

17. "Mold Allergy," Mayo Clinic, June 21, 2021, https://www.mayoclinic.org/diseases -conditions/mold-allergy/symptoms-causes/syc-20351519.

18. Ibid.

19. P. K. Daniel, "How Climate Change Impacts Allergies and Asthma," Allergy & Asthma Network, https://allergyasthmanetwork.org/news/something-the-air/.

20. Gennaro D'Amato et al., "Meteorological Conditions, Climate Change, New Emerging Factors, and Asthma and Related Allergic Disorders: A Statement of the World Allergy Organization," *World Allergy Organization Journal* 8, no. 1 (2015): 25, https://www.ncbi.nlm.nih.gov/pmc/articles/PMC4499913/.

21. Jennifer M. Albertine et al., "Projected Carbon Dioxide to Increase Grass Pollen and Allergen Exposure Despite Higher Ozone Levels," *PLoS One* 9, no. 11 (2014): e111712, https://www.ncbi.nlm.nih.gov/pmc/articles/PMC4221106/.

22. Tia Ghose, "Climate Change May Worsen Mold Allergies," Live Science, December 10, 2013, https://www.livescience.com/41838-climate-change-worsens-allergies.html.

23. Paul Gabrielsen, "Yes, Allergy Seasons Are Getting Worse. Blame Climate Change," University of Utah, February 8, 2021, https://attheu.utah.edu/facultystaff/pollen-seasons/.

24. Ibid.

25. "Pollen Counts Defined," American Academy of Allergy, Asthma & Immunology, https://www.aaaai.org/Tools-for-the-Public/Allergy,-Asthma-Immunology-Glossary/Pollen-Counts-Defined.

26. Ibid.

27. "How Does Rain Affect Pollen Levels?" Asthma and Allergy Foundation of America, https://community.aafa.org/blog/how-does-rain-affect-pollen-levels.

28. Bert Markgraf, "Does Rain Raise or Lower the Pollen Count?" Sciencing, August 24, 2018, https://sciencing.com/rain-raise-lower-pollen-count-23009.html.

29. Janet French, "10 Biggest Pollen Allergy Questions," *Allergic Living*, April 16, 2015, https://www.allergicliving.com/2015/04/16/10-biggest-pollen-allergy-questions/.

30. "What Causes or Triggers Asthma?" Asthma and Allergy Foundation of America, https://www.aafa.org/asthma-triggers-causes/.

31. Brian P. Dunleavy, "Thunderstorms Linked to 3,000 ER Visits a Year in Seniors with Asthma, COPD," UPI, August 10, 2020, https://www.upi.com/Health

_News/2020/08/10/Thunderstorms-linked-to-3000-ER-visits-a-year-in-seniors
-with-asthma-COPD/7411597065119/.

32. "Warming Seas May Increase Frequency of Extreme Storms," NASA Global Climate
Change, January 28, 2019, https://climate.nasa.gov/news/2837/warming-seas-may
-increase-frequency-of-extreme-storms/.

33. "National Climate Report—Annual 2020," National Centers for Environmental
Information, January 2021, https://www.ncdc.noaa.gov/sotc/national/202013.

34. Anthony LeRoy Westerling, "Wildfires in West Have Gotten Bigger, More Frequent,
and Longer Since the 1980s," *Conversation*, May 23, 2016, https://theconversation.
com/wildfires-in-west-have-gotten-bigger-more-frequent-and-longer-since-the
-1980s-42993.

35. Jessica Merzdorf, "A Drier Future Sets the Stage for More Wildfires," NASA Global
Climate Change, July 9, 2019, https://climate.nasa.gov/news/2891/a-drier-future
-sets-the-stage-for-more-wildfires/.

36. "Satellite Sees Smoke from Siberian Fires Reach the US Coast," NASA, June 11,
2012, https://www.nasa.gov/mission_pages/fires/main/siberia-smoke.html.

37. Cedars-Sinai Staff, "How Does Wildfire Smoke Affect Your Health?" Cedars-
Sinai, August 30, 2019, https://www.cedars-sinai.org/blog/smoke-from-wildfires.
html.

38. "What Is a Firestorm?" SciJinks, December 15, 2020, https://scijinks.gov/firestorm/.

39. Bob Berwyn, "How Wildfires Can Affect Climate Change (and Vice Versa)," Inside
Climate News, August 23, 2018, https://insideclimatenews.org/news/23082018
/extreme-wildfires-climate-change-global-warming-air-pollution-fire-management
-black-carbon-co2/.

40. "Wildfires and Indoor Air Quality (IAQ)," Environmental Protection Agency, https://
www.epa.gov/indoor-air-quality-iaq/wildfires-and-indoor-air-quality-iaq.

41. "Air Pollution," Asthma and Allergy Foundation of America, https://www.aafa.org
/air-pollution-smog-asthma/#:~:text=Ozone%20triggers%20asthma%20
because%20it,Ozone%20can%20reduce%20lung%20function.

42. "Climate Change Decreases the Quality of the Air We Breathe," Centers for Disease Control and Prevention, https://www.cdc.gov/climateandhealth/pubs/AIR-QUALITY-Final_508.pdf.

43. "Disparities in the Impact of Air Pollution," American Lung Association, April 20, 2020, https://www.lung.org/clean-air/outdoors/who-is-at-risk/disparities.

44. "Asthma Facts and Figures," Asthma and Allergy Foundation of America, https://www.aafa.org/asthma-facts/.

45. Jennifer Langston, "People of Color Exposed to More Pollution from Cars, Trucks, Power Plants during 10-Year Period," University of Washington News, September 14, 2017, https://www.washington.edu/news/2017/09/14/people-of-color-exposed-to-more-pollution-from-cars-trucks-power-plants-during-10-year-period/.

46. Ibid.

47. "Disparities in the Impact of Air Pollution."

48. "Asthma and Hispanic Americans," Office of Minority Health, February 11, 2021, https://minorityhealth.hhs.gov/omh/browse.aspx?lvl=4&lvlid=60.

49. Rubi Castillo et al., "Prevalence of Asthma Disparities amongst African-American Children," SMYSP, https://www.columbus.gov/uploadedFiles/Public_Health/Content_Editors/Community_Health/Minority_Health/Asthma%20dispatries%20amongest%20African%20American%20Children.pdf.

50. "Asthma Facts and Figures."

51. "Asthma and African Americans," Office of Minority Health, February 11, 2021, https://minorityhealth.hhs.gov/omh/browse.aspx?lvl=4&lvlid=15.

52. "Asthma Triggers: Gain Control," Environmental Protection Agency, https://www.epa.gov/asthma/asthma-triggers-gain-control.

53. Mayo Clinic Staff, "Childhood Asthma," Mayo Clinic, https://www.mayoclinic.org/diseases-conditions/childhood-asthma/symptoms-causes/syc-20351507#:~:text=Common%20childhood%20asthma%20signs%20and,Shortness%20of%20breath.

54. "What Is Indoor Air Quality?" Asthma and Allergy Foundation of America, https://www.aafa.org/indoor-air-quality/.

55. "You Can Control Mold," Centers for Disease Control and Prevention, https://www .cdc.gov/mold/control_mold.htm.

Chapter Six:
Shining Light on Winter Blues

1. Mayo Clinic Staff, "Depression in Women: Understanding the Gender Gap," Mayo Clinic, https://www.mayoclinic.org/diseases-conditions/depression/in-depth/depres sion/art-20047725.

2. Dr. Norman Rosenthal, "How Not to Get Seasonal Affective Disorder," *Daily Mail*, October 26, 2019, https://www.dailymail.co.uk/health/article-7616561/DR-NORMAN -ROSENTHAL-reveals-not-Seasonal-Affective-Disorder-winter.html.

3. Craig Newnes, *Inscription, Diagnosis, Deception and the Mental Health Industry* (London, Palgrave Macmillan, 2016), http://link.springer.com/content/pdf/10.1057%2F9781 137312969.pdf.

4. Rosenthal, "How Not to Get Seasonal Affective Disorder."

5. "Seasonal Affective Disorder," National Institute of Mental Health, https://www .nimh.nih.gov/health/publications/seasonal-affective-disorder/index.shtml.

6. "How to Know If You Have Seasonal Affective Disorder with Kelly Rohan, PhD," video, American Psychological Association, January 29, 2020, https://www.youtube .com/watch?v=3hvVfw0FzfY&t=1869s.

7. "Speaking of Psychology: How to Know If You Have Seasonal Affective Disorder," American Psychological Association, https://www.apa.org/research/action/speaking -of-psychology/seasonal-affective-disorder.

8. "Seasonal Affective Disorder."

9. J. J. Wurtman, "Carbohydrate Craving: Relationship between Carbohydrate Intake and Disorders of Mood," *Drugs* 39, Supplement 3 (1990): 49–52, https://pubmed .ncbi.nlm.nih.gov/2197075/.

10. "Seasonal Affective Disorder."

11. "Essential Guide to Serotonin and the Other Happy Hormones in Your Body," blog post, Atlas Biomed, May 27, 2020, https://atlasbiomed.com/blog/serotonin-and-other-happy-molecules-made-by-gut-bacteria/.

12. "Seasonal Affective Disorder."

13. Erin Bailey, "Experts and Students Discuss Seasonal Affective Disorder at Yale," *Yale Daily News*, November 29, 2020, https://yaledailynews.com/blog/2020/11/29/experts-and-students-discuss-seasonal-affective-disorder-at-yale/.

14. Sherri Melrose, "Seasonal Affective Disorder: An Overview of Assessment and Treatment Approaches," *Depression Research and Treatment* 2015 (2015): 178564, https://www.ncbi.nlm.nih.gov/pmc/articles/PMC4673349/.

15. "Hello Sunshine! Soaking Up in Sun Helps in Reducing Stress," India TV News, November 5, 2016, https://www.indiatvnews.com/lifestyle/news-soaking-up-in-sun-helps-in-reducing-stress-355355.

16. Alec Sears, "BYU Psychologist, Physicist, and Statistician Collaborate on Unique Study," Brigham Young University News, November 2, 2016, https://news.byu.edu/news/byu-psychologist-physicist-and-statistician-collaborate-unique-study.

17. Jon McBride, "Sunshine Matters a Lot to Mental Health," Brigham Young University Physics and Astronomy, https://physics.byu.edu/department/news/50.

18. "Seasonal Affective Disorder."

19. Ibid.

20. Dr. Zahid Naeem, "Vitamin D Deficiency: An Ignored Epidemic," *International Journal of Health Sciences* 4, no. 1 (January 2010): V–VI, https://www.ncbi.nlm.nih.gov/pmc/articles/PMC3068797/.

21. Sue Penckofer, PhD, RN, et al., "Vitamin D and Depression: Where Is All the Sunshine?" *Issues in Mental Health Nursing* 31, no. 6 (June 2010): 385–93, https://www.ncbi.nlm.nih.gov/pmc/articles/PMC2908269/.

22. Ibid.

23. "Vitamin D Deficiency, Depression Linked in UGA-Led International Study," UGA Today, December 2, 2014, https://news.uga.edu/vitamin-d-deficiency-depression/.

24. Ibid.

25. "Vitamin D Deficiency Linked to Mental Health," UGA Today, January 18, 2015, https://news.uga.edu/vitamin-d-deficiency-linked-to-mental-health/.

26. Jessica Blaszczak, "10 Things You Didn't Know About Seasonal Affective Disorder," PsychCentral, May 17, 2016, https://psychcentral.com/lib/10-things-you-dont-know-about-seasonal-affective-disorder.

27. Claudio N. Soares and Brook Zitek, "Reproductive Hormone Sensitivity and Risk for Depression across the Female Life Cycle: A Continuum of Vulnerability?" *Journal of Psychiatry & Neuroscience* 33, no. 4 (July 2008): 331–43, https://www.ncbi.nlm.nih.gov/pmc/articles/PMC2440795/.

28. J. M. Eagles, "Seasonal Affective Disorder: A Vestigial Evolutionary Advantage?" *Medical Hypotheses* 63, no. 5 (2004): 767–72, https://pubmed.ncbi.nlm.nih.gov/15488644/.

29. Ibid.

30. "Depression," National Institute of Mental Health, February 2018, https://www.nimh.nih.gov/health/topics/depression/index.shtml.

31. Simon N. Young, "How to Increase Serotonin in the Human Brain Without Drugs," *Journal of Psychiatry & Neuroscience* 32, no. 6 (November 2007): 394–99, https://www.ncbi.nlm.nih.gov/pmc/articles/PMC2077351/.

32. Mark Mikulka, "The Ultimate Guide to Light Measurement," Lumitex, June 19, 2018, https://www.lumitex.com/blog/light-measurement.

33. Carla Lanca et al., "The Effects of Different Outdoor Environments, Sunglasses, and Hats on Light Levels: Implications for Myopia Prevention," *Translational Vision Science & Technology* 8, no. 4 (July 2019): 7, https://www.ncbi.nlm.nih.gov/pmc/articles/PMC6656201/.

34. Mathias Adamsson, Thorbjörn Laike, and Takeshi Morita, "Annual Variation in Daily Light Exposure and Circadian Change of Melatonin and Cortisol Concentrations at a Northern Latitude with Large Seasonal Differences in Photoperiod Length," *Journal of Physiological Anthropology* 36 (2017): 6, https://www.ncbi.nlm.nih.gov/pmc/articles/PMC4952149/.

35. Young, "How to Increase Serotonin in the Human Brain."

36. "Seasonal Affective Disorder."

37. Christopher Klein, "8 Things You May Not Know About Daylight Saving Time," History.com, October 26, 2020, https://www.history.com/news/8-things-you-may-not-know-about-daylight-saving-time.

38. Ibid.

39. Matthew J. Kotchen and Laura E. Grant, "Does Daylight Saving Time Save Energy? Evidence from a Natural Experiment in Indiana," National Bureau of Economic Research, October 2008, https://www.nber.org/papers/w14429.

40. "Daylight Saving Time: 4 Tips to Help your Body Adjust," Cleveland Clinic, February 28, 2020, https://health.clevelandclinic.org/daylight-savings-time-change-4-tips-to-help-your-body-adjust/.

41. Melinda A. Ma and Elizabeth H. Morrison, "Neuroanatomy, Nucleus Suprachiasmatic," *StatPearls* (July 31, 2020), https://www.ncbi.nlm.nih.gov/books/NBK546664/.

42. "Seasonal Affective Disorder (SAD)," American Psychiatric Association, October 2020, https://www.psychiatry.org/patients-families/depression/seasonal-affective-disorder.

43. Ibid.

44. Kelly J. Rohan et al., "Cognitive-Behavioral Therapy vs. Light Therapy for Preventing Winter Depression Recurrence: Study Protocol for a Randomized Controlled Trial," *Trials* 14 (2013): 82, https://www.ncbi.nlm.nih.gov/pmc/articles/PMC3652773/.

45. "Seasonal Affective Disorder: More Than the Winter Blues," American Psychological Association, 2014, https://www.apa.org/topics/depression/seasonal-affective-disorder.

46. "Hippocrates Media: Winter Blues—Seasonal Affective Disorder," Facebook post, October 9, 2016, https://www.facebook.com/303503566446606/posts/925419130921710/.

47. Norman Rosenthal, MD, "A Sneak Peek at the Paperback Version of *Transcendence*," blog post, July 27, 2012, https://www.normanrosenthal.com/blog/2012/07/sneak-peak-transcendence-paperback/.

48. "Transcendental Meditation Technique," Wikipedia, June 21, 2021, https://en
.wikipedia.org/wiki/Transcendental_Meditation_technique.

49. Sharon, "The Ultimate Mindfulness Meditation Guide for Beginners," Learn Relaxation Techniques, https://learnrelaxationtechniques.com/how-to-do-transcendental
-meditation-step-by-step/.

50. Martin Luenendonk, "These Are the 10 Most Exciting Mantras for Meditation,"
Cleverism, September 25, 2019, https://www.cleverism.com/mantras-for-meditation/.

51. "Mantra Meaning," Soul Yoga, https://soulyoga.teachable.com/courses/monthly
-mantra/lectures/13420233.

52. Sharon, "The Ultimate Mindfulness Meditation Guide for Beginners."

Chapter Seven:
Weathering Autoimmune Flare-Ups

1. Tessa Love, "Why Are Autoimmune Diseases on the Rise?" Elemental, April 10, 2019,
https://elemental.medium.com/autoimmunity-is-a-disorder-of-our-time-a7f1c45d6907.

2. "Autoimmune Disease . . ." American Autoimmune Related Diseases Association, Inc.,
https://www.aarda.org/news-information/statistics/.

3. "Progress in Autoimmune Diseases Research," National Institutes of Health, March
2005, https://www.niaid.nih.gov/sites/default/files/adccfinal.pdf.

4. Carly Ray and Xue Ming, "Climate Change and Human Health: A Review of
Allergies, Autoimmunity, and the Microbiome," *International Journal of Environmental
Research and Public Health* 17, no. 13 (July 4, 2020): 4814, https://pubmed.ncbi.nlm
.nih.gov/32635435/.

5. Emily Sheng, "Did Climate-Changing Pollution Give Me Lupus?" *Tampa Bay Times*,
November 22, 2019, https://www.tampabay.com/opinion/2019/11/22/did-climate
-changing-pollution-give-me-lupus-column/.

6. Chan-Na Zhao et al., "Emerging Role of Air Pollution in Autoimmune Diseases,"
Autoimmunity Reviews 18, no. 6 (June 2019): 607–14, https://pubmed.ncbi.nlm.nih
.gov/30959217/.

7. Jayson MacLean, "Air Pollution and Autoimmune Diseases Are Linked, Says Canadian Study," Cantech Letter, January 9, 2016, https://www.cantechletter.com/2016/01/air-pollution-and-autoimmune-disease-linked-says-canadian-study/.

8. "Doctor S. M. Akerkar, Consultant Rheumatologist, Mumbai," Arthritis Support Board, http://www.arthritissupportboard.com/AboutDrAkerkar.aspx.

9. Dr. S. M. Akerkar, "How Does Weather Affect Rheumatoid Arthritis?" blog post, Arthritis Support Board, September 4, 2011, https://doctorakerkar.wordpress.com/2011/09/04/how-does-weather-affect-rheumatoid-arthritis/.

10. "What Is Arthritis?" Arthritis Foundation, https://www.arthritis.org/health-wellness/about-arthritis/understanding-arthritis/what-is-arthritis.

11. Homepage, Cloudy With a Chance of Pain, https://www.cloudywithachanceofpain.com/.

12. Erik Timmermans et al., "The Influence of Weather Conditions on Joint Pain in Older People with Osteoarthritis: Results from the European Project on Osteoarthritis," Journal of Rheumatology 42, no. 10 (September 2015), https://www.researchgate.ne/publication/281518431_The_Influence_of_Weather_Conditions_on_Joint_Pain_in_Older_People_with_Osteoarthritis_Results_from_the_European_Project_on_OSteoArthritis.

13. Dr. Rothbard, "Cold Weather's Impact on Autoimmune Disease Flares," American Autoimmune Related Diseases Association, Inc., October 25, 2013, https://www.aarda.org/cold-weather-impact-autoimmune-disease-flares/.

14. Ibid.

15. Charlotte Hilton Andersen, "Cold Weather Joint Pain: 15 Tips for Managing It," Creaky Joints, November 22, 2019, https://creakyjoints.org/living-with-arthritis/cold-weather-joint-pain-arthritis/.

16. Ibid.

17. Mayo Clinic Staff, "Psoriasis," Mayo Clinic, https://www.mayoclinic.org/diseases-conditions/psoriasis/symptoms-causes/syc-20355840.

18. "Layers of the Skin," National Cancer Institute, https://www.training.seer.cancer.gov/melanoma/anatomy/layers.html.

19. "About Psoriasis," National Psoriasis Foundation, https://www.psoriasis.org/about-psoriasis/.

20. "Psoriasis Statistics," National Psoriasis Foundation, https://www.psoriasis.org/psoriasis-statistics/.

21. "Can Weather Affect Your Psoriasis and Psoriatic Arthritis?" WebMD, May 20, 2021, https://www.webmd.com/arthritis/psoriatic-arthritis/psa-weather-changes.

22. "Active and Mindful Lifestyles," National Psoriasis Foundation, https://www.psoriasis.org/active-and-mindful-lifestyles/.

23. Varsha Vats, "World Psoriasis Day 2020: Follow These Expert Recommended Tips to Manage Psoriasis during Upcoming Winter," NDTV, October 29, 2020, https://www.ndtv.com/health/world-psoriasis-day-follow-these-expert-recommended-tips-to-manage-psoriasis-during-upcoming-winter-2317456.

24. "Thawing the Fact from Fiction behind Winter Flares," National Psoriasis Foundation, https://www.psoriasis.org/advance/thawing-the-fact-from-fiction-behind-winter-flares/.

25. "Phototherapy," National Psoriasis Foundation, https://www.psoriasis.org/phototherapy/.

26. "Does Light Therapy (Phototherapy) Help Reduce Psoriasis Symptoms?" InformedHealth.org, May 18, 2017, https://www.ncbi.nlm.nih.gov/books/NBK435696/.

27. K. Rappersberger et al., "Photosensitivity and Hyperpigmentation in Amiodarone-Treated Patients: Incidence, Time Course, and Recovery," *Journal of Investigative Dermatology* 93, no. 2 (August 1989): 201–9, https://pubmed.ncbi.nlm.nih.gov/2754275/.

28. Sara Coughlin, "What Really Causes the Butterfly Rash in Lupus—and What to Do About It," Health.com, March 15, 2019, https://www.health.com/condition/lupus/lupus-butterfly-rash.

29. "UV Exposure: What You Need to Know," Lupus Foundation of America, June 7, 2021, https://www.lupus.org/resources/uv-exposure-what-you-need-to-know.

30. "Unlocking the Reasons Why Lupus Is More Common in Women," Lupus Foundation of America, https://www.lupus.org/resources/unlocking-the-reasons-why-lupus-is-more-common-in-women.

31. "African Americans and Lupus," Lupus Foundation of America, 2013, https://www
   .lupus.org/s3fs-public/Doc%20-%20PDF/Ohio/African%20Americans%20and%20
   Lupus.pdf.

32. Michelle Brandt, "Research Shows Why Lupus May Be More Common in Black,
   Asian People," Scope, April 13, 2010, https://scopeblog.stanford.edu/2010/04/13
   /lupus_finding/.

33. "Health Disparities in SLE," Lupus Initiative, https://thelupusinitiative.org/slides
   /pdf/PP_HealthDisparities.pdf.

34. Wiley-Blackwell, "Oral Contraceptives Associated with Increased Risk of
   Lupus," ScienceDaily, April 8, 2009, https://www.sciencedaily.com/releases/2009
   /04/090407130912.htm.

35. University of Cincinnati Academic Health Center, "Environmental Factors May
   Trigger Lupus Onset, Progression," ScienceDaily, November 16, 2017, https://www
   .sciencedaily.com/releases/2017/11/171116142056.htm.

36. Katherine Lee, "Study Shows Changes in Weather, Environment May Indeed Affect
   Your Lupus Symptoms," Everyday Health, November 10, 2019, https://www
   .everydayhealth.com/lupus/study-shows-changes-in-weather-environment-may
   -indeed-affect-your-lupus-symptoms/.

37. Elisa Couto Gomes and Geraint Florida-James, "Lung Inflammation, Oxidative Stress,
   and Air Pollution," IntechOpen.com, May 14, 2014, https://www.intechopen.com
   /books/lung-inflammation/lung-inflammation-oxidative-stress-and-air-pollution.

38. Marlene Cimons, "Why Rising Temperatures Could Make Life Harder for
   Lupus Patients," PBS, December 3, 2019, https://www.pbs.org/wnet/peril-and
   -promise/2019/12/extreme-weather-will-hurt-lupus-patients/.

39. "10 Truths about UV Radiation," Lupus Foundation of America, https://www.lupus
   .org/resources/10-truths-about-uv-radiation.

Chapter Eight:
Heat, Sleep, and Memory

1. Jose Guillermo Cedeño Laurent et al., "Reduced Cognitive Function during a Heat Wave among Residents of Non-Air-Conditioned Buildings: An Observational Study of Young Adults in the Summer of 2016," *PLoS Medicine* 15, no. 7 (July 2018): e1002605, https://www.ncbi.nlm.nih.gov/pmc/articles/PMC6039003/.

2. Björn Rasch and Jan Born, "About Sleep's Role in Memory," *Physiological Reviews* 93, no. 2 (April 2013): 681–766, https://www.ncbi.nlm.nih.gov/pmc/articles /PMC3768102/.

3. University of Bern, "'Goldilocks' Neurons Promote REM Sleep," ScienceDaily, June 19, 2019, https://www.sciencedaily.com/releases/2019/06/190619111248.htm.

4. Richard Boyce and Antoine Adamantidis, "REM Sleep on It!" *Neuropsychopharmacology* 42, no. 1 (January 2017): 375, https://www.ncbi.nlm.nih.gov/pmc/articles/PMC5143514/.

5. "Sleep On It: How Snoozing Strengthens Memories," NIH News in Health, April 2013, https://newsinhealth.nih.gov/2013/04/sleep-it.

6. Ibid.

7. Danielle Pacheco, "Memory and Sleep," Sleep Foundation, November 13, 2020, https://www.sleepfoundation.org/how-sleep-works/memory-and-sleep.

8. Brian Kahn, "Here's How Much US Summers Have Warmed Since 1970," Climate Central, June 3, 2014, https://www.climatecentral.org/news/us-summer -temperatures-climate-change-17510.

9. Inga Kiderra, "How Climate Change Makes Us All Lose Sleep," University of California, June 5, 2017, https://www.universityofcalifornia.edu/news/how-climate -change-makes-us-all-lose-sleep.

10. Ibid.

11. Ibid.

12. Alan Miller et al., "Chilling Prospects: Providing Sustainable Cooling for All," Sustainable Energy for All, https://www.seforall.org/sites/default/files/SEforALL _CoolingForAll-Report_0.pdf.

13. Kiderra, "How Climate Change Makes Us All Lose Sleep."

14. "Sleep and Sleep Disorder: Data and Statistics," Centers for Disease Control and Prevention, May 2, 2017, https://www.cdc.gov/sleep/data_statistics.html.

15. "Optic Chiasma," Healthline, January 21, 2018, https://www.healthline.com/human-body-maps/optic-chiasm#2.

16. "Take a Warm Bath 1–2 Hours Before Bedtime to Get Better Sleep, Researchers Find," UT News, July 19, 2019, https://news.utexas.edu/2019/07/19/take-a-warm-bath-1-2-hours-before-bedtime-to-get-better-sleep-researchers-find/.

17. Shahab Haghayegh, "Want Better Sleep? Science Proves a Bath 1–2 Hours Before Bed Will Improve Your Sleep," Thrive Global, August 5, 2019, https://thriveglobal.com/stories/want-better-sleep-science-proves-a-bath-1-2-hours-before-bed-will-improve-your-sleep/.

18. Terry Gross, "How the 'Lost Art' of Breathing Can Impact Sleep and Resilience," NPR, May 27, 2020, https://www.npr.org/sections/health-shots/2020/05/27/862963172/how-the-lost-art-of-breathing-can-impact-sleep-and-resilience.

19. Joel Jean, "I'm Living in a Carbon Bubble. Literally," Medium, December 3, 2016, https://medium.com/@joeljean/im-living-in-a-carbon-bubble-literally-b7c391e8ab6.

20. P. Strøm-Tejsen et al., "The Effects of Bedroom Air Quality on Sleep and Next-Day Performance," *Indoor Air* 26, no. 5 (October 2016): 679–86, https://onlinelibrary.wiley.com/doi/full/10.1111/ina.12254#ina12254-bib-0015.

21. Kathryn M. Orzech et al., "Digital Media Use in the 2 h Before Bedtime Is Associated with Sleep Variables in University Students," *Computers in Human Behavior* 55, no. A (February 2016): 43–50, https://www.ncbi.nlm.nih.gov/pmc/articles/PMC5279707/.

Chapter Nine:
Food Choices and Carbon Footprints

1. Phoebe Tran, "Food Biodiversity: Where Flavor and Sustainability Meet," Food + Tech Connect, February 4, 2019, https://foodtechconnect.com/2019/02/04/food-biodiversity-flavor-sustainability-meet/.

2. Cheikh Mbow et al., "Food Security," chapter 5 of the Special Report on Climate Change and Land, Intergovernmental Panel on Climate Change, https://www.ipcc.ch/srccl/chapter/chapter-5/.

3. Agrilinks Team, "Linking Biodiversity Targets to Food System Sustainability: The Agrobiodiversity Index," Agrilinks, January 29, 2020, https://www.agrilinks.org/post/linking-biodiversity-targets-food-system-sustainability.

4. Tran, "Food Biodiversity."

5. Agrilinks Team, "Linking Biodiversity Targets to Food System Sustainability."

6. "Natural Resources for Food System Activities," FutureLearn, https://www.futurelearn.com/info/courses/food-systems-southeast-asia/0/steps/83755.

7. "The Disappearance of Biodiversity Crucial for Food & Agriculture," Organic Without Boundaries, March 22, 2019, https://www.organicwithoutboundaries.bio/2019/03/22/biodiversity-food-agriculture/.

8. Tariq Khokhar, "Chart: Globally, 70% of Freshwater Is Used for Agriculture," World Bank Blogs, March 22, 2017, https://blogs.worldbank.org/opendata/chart-globally-70-freshwater-used-agriculture.

9. "UN Report Suggests Humans Should Adopt Plant-Based Diets to Fight Climate Change," Vegan Food & Living, August 8, 2019, https://www.veganfoodandliving.com/news/un-report-says-humans-must-adopt-plant-based-diets-to-fight-climate-change/.

10. "Land Is a Critical Resource, IPCC Report Says," Intergovernmental Panel on Climate Change, August 8, 2019, https://www.ipcc.ch/2019/08/08/land-is-a-critical-resource_srccl/.

11. "Working Group II Impacts, Adaptation, and Vulnerability," Intergovernmental Panel on Climate Change, https://www.ipcc.ch/working-group/wg2/.

12. "Peoples' Climate Vote: Results," United Nations Development Programme, https://www.undp.org/content/dam/undp/library/km-qap/UNDP-Oxford-Peoples-Climate-Vote-Results.pdf.

13. Ibid.

14. "Earth Day Network and Yale Program on Climate Change Communication Present New Research on Consumer Food Habits Related to Climate Change," Yale Program on Climate Change Communication, February 17, 2020, https://climatecommunication .yale.edu/news-events/earth-day-network-and-yale-program-on-climate-change -communication-present-new-research-on-consumer-food-habits-related-to-climate -change/.

15. Kathryn Asher et al., "Study of Current and Former Vegetarians and Vegans," Humane Research Council, 2014, https://faunalytics.org/wp-content/uploads/2015/06 /Faunalytics_Current-Former-Vegetarians_Full-Report.pdf.

16. Frank Mitloehner, "The Bogus Burger Blame," CLEAR Center, February 12, 2021, https://clear.ucdavis.edu/blog/bogus-burger-blame.

17. "Sources of Greenhouse Gas Emissions," Environmental Protection Agency, https:// www.epa.gov/ghgemissions/sources-greenhouse-gas-emissions.

18. Sara Place, "Gassy Cows? Facts About Beef's Carbon Emissions," GreenBiz, May 14, 2018, https://www.greenbiz.com/article/gassy-cows-facts-about-beefs-carbon-emissions -sponsored.

19. "Greenhouse Gases (GHGs) and the US Dairy Industry," https://saylordotorg.github .io/text_the-sustainable-business-case-book/s14-03-greenhouse-gases-ghgs-and -the-.html.

20. "Sources of Greenhouse Gas Emissions."

21. Ibid.

22. Henry Fountain, "Belching Cows and Endless Feedlots: Fixing Cattle's Climate Issues," New York Times, October 21, 2020, https://www.nytimes.com/2020/10/21 /climate/beef-cattle-methane.html.

23. Quirin Schiermeier, "Global Methane Levels Soar to Record High," Nature, July 14, 2020, https://www.nature.com/articles/d41586-020-02116-8.

24. "Global Livestock Environmental Assessment Model (GLEAM)," Food and Agriculture Organization of the United Nations, http://www.fao.org/gleam/results/en/.

25. "17 Organizations Promoting Regenerative Agriculture Around the Globe," FoodTank, May 2018, https://foodtank.com/news/2018/05/organizations-feeding-healing-world-regenerative-agriculture-2/.

26. Kyra Blank, "Regenerative Ranching Could Solve Climate Change," Regeneration International, September 20, 2020, https://regenerationinternational.org/2020/09/20/regenerative-ranching-could-solve-climate-change/.

27. Richard Gray, "Why Soil Is Disappearing from Farms," BBC, https://www.bbc.com/future/bespoke/follow-the-food/why-soil-is-disappearing-from-farms/.

28. Jeremy Hobson, "UN Report Links Soil Degradation to Climate Change," WBUR, September 20, 2019, https://www.wbur.org/hereandnow/2019/09/20/soil-degradation-climate-change.

29. Adele Peters, "Is It Possible to Raise a Carbon-Neutral Cow?" *Fast Company*, July 24, 2019, https://www.fastcompany.com/90368127/is-it-possible-to-raise-a-carbon-neutral-cow.

30. Barry Estabrook, "This Man Wants You to Eat More Meat," EW Content, http://ew.content.allrecipes.com/article/290723/this-man-wants-you-to-eat-more-meat.

31. Gabriel Popkin, "Can 'Carbon Smart' Farming Play a Key Role in the Climate Fight?" Yale Environment 360, March 31, 2020, https://e360.yale.edu/features/can-carbon-smart-farming-play-a-key-role-in-the-climate-fight.

32. Meg Wilcox, "To Meet Ambitious Emissions Goals, Large Food Companies Are Looking to Lock Carbon in Soil," *Smithsonian Magazine*, February 19, 2021, https://www.smithsonianmag.com/innovation/meet-ambitious-emissions-goals-large-food-companies-are-looking-lock-carbon-soil-180977053/.

33. Cortney Ahern Renton, Claire Huntley Lafave, and Katie Sierks, "Farmers on the Front Lines of the Regenerative Agriculture Transition," Conservation Finance Network, April 15, 2020, https://conservationfinancenetwork.org/2020/04/15/farmers-on-the-frontlines-of-the-regenerative-agriculture-transition.

34. Vernon Graham, "Australian Cattle Industry Has Cut Its Greenhouse Gas Emissions in Half," Farmonline, May 21, 2019, https://www.farmonline.com.au/story/6135115/no-bull-beef-cattle-industry-is-cutting-its-carbon-footprint/.

35. Angus Verley et al., "Carbon Neutral Livestock Production—Consumers Want It and Farmers Say It Is Achievable," ABC News (Australia), June 7, 2019, https://www.abc.net.au/news/rural/2019-06-08/carbon-neutral-livestock-achievable-by-2030-says-mla/11046592.

36. "The Global Staple," Ricepedia, http://ricepedia.org/rice-as-food/the-global-staple-rice-consumers.

37. "Climate Change Scientist Joins the Columbia Mailman School," Public Health Now, August 23, 2019, https://www.publichealth.columbia.edu/public-health-now/news/climate-change-scientist-joins-columbia-mailman-school.

38. Shannon Gupta, "Climate Change Is Hurting US Corn Farmers—and Your Wallet," CNN Money, April 21, 2017, https://money.cnn.com/2017/04/20/news/corn-farmers-climate-change/index.html.

39. "Derecho 2020 Path," Healthy Weight Research Network, December 27, 2020, https://hwrn.org/4qrfe1o/derecho-2020-path-909ddc.

40. Natalina Sents, "How Much Did the Derecho Damage Iowa Agriculture?" Successful Farming, August 21, 2020, https://www.agriculture.com/news/crops/how-much-did-the-derecho-damage-iowa-agriculture.

41. "Extreme Weather Gets a Boost from Climate Change," Environmental Defense Fund, https://www.edf.org/climate/climate-change-and-extreme-weather.

42. "Climate and . . ." Climate.gov, https://www.climate.gov/news-features/department/climate-and?page=1.

43. Laura Fornero, "How to Reduce the Environmental Impact of Your Coffee Habit," Perfect Daily Grind, May 27, 2019, https://perfectdailygrind.com/2019/05/how-to-reduce-the-environmental-impact-of-your-coffee-habit/.

44. Judy Fleisher, "The Effect of Altitude on Coffee Flavor," Scribblers Coffee Co., October 16, 2017, https://scribblerscoffee.com/blogs/news/the-effect-of-altitude-on-coffee-flavor.

45. "Bringing Children from Isolation to Education," Project Alianza, https://www.projectalianza.org/why-the-coffeelands.

46. Scott, "How Does Elevation Affect the Taste of Coffee?" Driftaway Coffee, May 23, 2015, https://driftaway.coffee/elevation/.

47. "The Water Footprint of Your Coffee," Catholic Relief Services: Coffeelands, September 11, 2012, https://coffeelands.crs.org/2012/09/302-the-water-footprint-of-your-coffee/.

48. Fábio M. DaMatta et al., "Why Could the Coffee Crop Endure Climate Change and Global Warming to a Greater Extent Than Previously Estimated?" *Climatic Change* 152 (2019): 167–78, https://link.springer.com/article/10.1007/s10584-018-2346-4.

49. "Coffee," PlantVillage, https://plantvillage.psu.edu/topics/coffee/infos.

50. Bonnie Rochman, "Coffee Farming Adapts in the Face of Climate Change," Starbucks Stories, June 21, 2018, https://stories.starbucks.com/stories/2018/coffee-farming-adapts-in-the-face-of-climate-change/.

51. Oriana Ovalle Rivera, "Projected Shifts in *Coffea arabica* Suitability among Major Global Producing Regions Due to Climate Change," *PLoS One* 10, no. 4 (April 2015): e0124155, https://www.researchgate.net/publication/274699687_Projected_Shifts_in_Coffea_arabica_Suitability_among_Major_Global_Producing_Regions_Due_to_Climate_Change.

52. "Fairtrade Certified Coffee," Fairtrade America, https://www.fairtradeamerica.org/shop-fairtrade/fairtrade-products/coffee/.

53. Amy Shoenthal, "What Exactly Is Fair Trade, and Why Should We Care?" *Forbes*, December 14, 2018, https://www.forbes.com/sites/amyschoenberger/2018/12/14/what-exactly-is-fair-trade-and-why-should-we-care/.

54. Jonathan A. Hare et al., "A Vulnerability Assessment of Fish and Invertebrates to Climate Change on the Northeast US Continental Shelf," *PLoS One* 11, no. 2 (2016): e0146756, https://journals.plos.org/plosone/article?id=10.1371/journal.pone.0146756.

55. Maurice Tamman, "Special Report—Ocean Shock: Lobster's Great Migration Sets Up Boom and Bust," Reuters, October 30, 2018, https://www.reuters.com/article/oceans-tide-lobster-idINKCN1N428J.

56. David Barstow, "Scientists Are Mystified by Deaths of Lobsters in Long Island Sound," *New York Times*, October 18, 1999, https://www.nytimes.com/1999/10/18/nyregion/scientists-are-mystified-by-deaths-of-lobsters-in-long-island-sound.html.

57. Emily Greenhalgh, "Climate & Lobsters," Climate.gov, October 6, 2016, https://www.climate.gov/news-features/climate-and/climate-lobsters.

58. Ibid.

59. Laura Poppick, "Why Is the Gulf of Maine Warming Faster Than 99% of the Ocean?" Eos, November 12, 2018, https://eos.org/features/why-is-the-gulf-of-maine-warming-faster-than-99-of-the-ocean.

60. Greenhalgh, "Climate & Lobsters."

61. "Understanding Sustainable Seafood," National Oceanic and Atmospheric Administration, https://www.fisheries.noaa.gov/insight/understanding-sustainable-seafood.

62. "National Consumer Guide," Monterey Bay Aquarium Seafood Watch, https://www.seafoodwatch.org/globalassets/sfw/pdf/guides/seafood-watch-national-guide.pdf.

63. Shelby Brown, "The 10 Most Instagrammed Foods and Drinks of All Time," Spoon University, https://spoonuniversity.com/lifestyle/the-10-most-instagrammed-foods-and-drinks-of-all-time.

64. "How Much Water Does It Take to Grow an Avocado?" Danwatch, https://old.danwatch.dk/en/undersogelseskapitel/how-much-water-does-it-take-to-grow-an-avocado/.

65. "Santa Ana Wind Conditions Affect Avocado Tree Water Use," California Avocado Commission, October 15, 2019, https://www.californiaavocadogrowers.com/articles/santa-ana-wind-conditions-affect-avocado-tree-water-use.

66. "Why Our Love for Avocados Is Not Sustainable," Facebook post, Living Forests India, February 11, 2020, https://www.facebook.com/permalink.php?id=101796587972393&story_fbid=135831454568906.

67. "5 of Your Favorite Foods Threatened by Climate Change," Rainforest Alliance, June 25, 2019, https://www.rainforest-alliance.org/articles/5-favorite-foods-youre-about-to-lose-to-climate-change.

68. Michon Scott, "Climate & Chocolate," Climate.gov, February 10, 2016, https://www.climate.gov/news-features/climate-and/climate-chocolate.

69. "Climate Change & Cacao Farmers . . . Recipe for Disaster??" Chocolate Class, May 3, 2019, https://chocolateclass.wordpress.com/2019/05/03/climate-change-cacao-farmers-recipe-for-disaster/.

70. Scott, "Climate & Chocolate."

71. "Sample Records for Cacao Criollo Porcelana," Science.gov, https://www.science
    .gov/topicpages/c/cacao+criollo+porcelana.

72. "Cocoa," Rainforest Alliance, https://www.rainforest-alliance.org/tags/chocolate.

73. "Farmers Market Talking Points," Farmers Market Coalition, 2016, https://
    farmersmarketcoalition.org/wp-content/uploads/2016/07/FMC_TalkingPoints
    _2016.pdf.

74. "Global Food Losses and Food Waste," Food and Agriculture Organization of the
    United Nations, 2011, http://www.fao.org/3/mb060e/mb060e00.htm.

75. "Fight Climate Change by Preventing Food Waste, World Wildlife Fund, https://
    www.worldwildlife.org/stories/fight-climate-change-by-preventing-food-waste.

76. Chad Frischmann, "Opinion: The Climate Impact of the Food in the Back of Your
    Fridge," *Washington Post*, July 31, 2018, https://www.washingtonpost.com/news
    /theworldpost/wp/2018/07/31/food-waste/.

77. "Chapter 5: Food Security," Intergovernmental Panel on Climate Change, 2019,
    https://www.ipcc.ch/site/assets/uploads/2019/08/2f.-Chapter-5_FINAL.pdf.

78. "Worldwide Food Waste," United Nations Environment Programme, https://www
    .unenvironment.org/thinkeatsave/get-informed/worldwide-food-waste.

79. Daisy Simmons, "A Brief Guide to the Impacts of Climate Change on
    Food Production," Yale Climate Connections, September 18, 2019, https://
    yaleclimateconnections.org/2019/09/a-brief-guide-to-the-impacts-of-climate
    -change-on-food-production/.

80. "Food Loss and Waste," US Food & Drug Administration, https://www.fda.gov
    /food/consumers/food-loss-and-waste.

81. "Tips to Reduce Food Waste," US Food & Drug Administration, https://www.fda
    .gov/food/consumers/tips-reduce-food-waste.

Chapter 10:
Sustainable Fitness

1. Becca Glasser-Baker, "Update: NYC Triathlon Canceled Due to Dangerous Heat," *Metro*, July 16, 2019, https://www.metro.us/update-nyc-triathlon-cancelled-due-to -dangerous-heat/.

2. Stephen Wade, "Tokyo's Summer Heat Forces Triathlon Test to Be Shortened," Associated Press, August 15, 2019, https://apnews.com/article/a9da516f 93244533842751820294a88c.

3. Laura Millan Lombrana, "Arctic Sea Ice Shrank to Record Lows in July," Bloomberg, August 6, 2020, https://www.msn.com/en-us/weather/topstories/arctic-sea-ice -shrank-to-record-lows-in-july/ar-BB17E5fJ.

4. Kelly Kimball, "The Last Great Race," *Foreign Policy*, April 27, 2020, https:// foreignpolicy.com/2020/04/27/iditarod-climate-change-alaska-last-great -race/#:~:text=In%202007%2C%20some%2024%20mushers,prematurely%20 for%20the%20same%20reasons.&text=A%20recent%20Arctic%20Report%20 Card,winter%20temperatures%20in%20Alaska's%20history.

5. "From Cricket to Yachting: Five Sports Events Disrupted by Bushfires," *Times of Malta*, January 14, 2020, https://timesofmalta.com/articles/view/from-cricket-to-yachting -five-sports-events-disrupted-by-bushfires.763428.

6. Jack M. Germain, "Fitness Products Shatter Online Sales Records during Lockdown," *E-Commerce Times*, July 20, 2020, https://www.ecommercetimes.com/story/86762.html.

7. "Meet Keyword Explorer," Moz, https://analytics.moz.com/pro/keyword-explorer /keyword/suggestions?locale=en-US&q=outdoor%20gym.

8. "Scarborough Southwest Adult Fitness Improvements," City of Toronto, August 18, 2020, https://www.toronto.ca/city-government/planning-development/construction -new-facilities/improvements-expansion-redevelopment/ward-20-scarborough -southwest-adult-fitness-improvements/.

9. Ed Wright, "Free Outdoor Gym Apparatus in Canton Looks Like Something Straight Out of *The Jetsons*," Hometownlife.com, November 30, 2020, https://www .hometownlife.com/story/life/community/observer/canton/2020/11/30 /futuristic-outdoor-gym-canton-township-funded-developer/6327765002/.

10. "Panthers and Lowe's Build Unique Outdoor Fitness Space at Veterans Park in Charlotte," Panthers.com, September 25, 2020, https://www.panthers.com/news /panthers-lowes-fitness-space-veterans-park.

11. Nick Obradovich and James H. Fowler, "Climate Change May Alter Human Physical Activity Patterns," *Nature Human Behaviour* 1 (2017): 0097, https://nickobradovich .com/climatechange/climate-change-may-alter-human-physical-activity-patterns/.

12. Sarah Klein, "This Is What Happens to Your Body When You Exercise," *Huffpost*, September 4, 2013, https://www.huffpost.com/entry/body-on-exercise-what -happens-infographic_n_3838293.

13. Ibid.

14. "The Process of Breathing," Lumen Learning, https://courses.lumenlearning.com /suny-ap2/chapter/the-process-of-breathing-no-content/.

15. Klein, "This Is What Happens to Your Body."

16. "What Is the Effect of Heat and Humidity on Athletic Performance?" blog post, Intermountain Helathcare, June 21, 2014, https://intermountainhealthcare.org /blogs/topics/sports-medicine/2014/06/what-is-the-effect-of-heat-and-humidity -on-athletic-performance/.

17. "How Hot Weather Can Affect Your Heart When You Exercise," Cleveland Clinic, July 19, 2018, https://health.clevelandclinic.org/how-hot-weather-can-affect-your -heart-when-you-exercise/.

18. "Your Lungs and Exercise," *Breathe* 12, no. 1 (March 2016): 97–100, https://www .ncbi.nlm.nih.gov/pmc/articles/PMC4818249/.

19. "How Hot Weather Can Affect Your Heart."

20. "Exercise-Related Heat Exhaustion," Johns Hopkins Medicine, https://www .hopkinsmedicine.org/health/conditions-and-diseases/exerciserelated-heat -exhaustion.

21. "Medical Alert! Climate Change Is Harming Our Health."

22. Lindsey Barton Straus, JD, "High School Football Players Most Prone to Heat Illness, CDC Says," MomsTeam.com, April 20, 2017, https://www.momsteam.com/heat-stroke/high-school-football-players-most-prone-heat-illness-CDC-study-says.

23. "Climate, Health, and Equity: A Policy Action Agenda," ClimateHealthAction.org, https://climatehealthaction.org/media/cta_docs/PAA%20Fact%20Sheet.pdf.

24. "Medical Alert! Climate Change Is Harming Our Health."

25. Ibid.

26. Susan Yeargin, PhD, ATC, "Do Children Handle Heat as Well as Adults?" MomsTeam.com, July 13, 2012, https://www.momsteam.com/health-safety/children-handle-heat-as-well-as-adults-studies-say.

27. "Heat Illness: Keeping Youth Sports Athletes Safe," TrueSport.org, May 28, 2018, https://truesport.org/hydration/heat-illness-youth-sports/.

28. Olivia Rosane, "Wildfire Smoke Is More Toxic Than Other Forms of Air Pollution, Study Finds," EcoWatch, March 8, 2021, https://www.ecowatch.com/wildfire-smoke-toxic-pollution-2650974582.html.

29. Jim Wilson, "Is It Safe to Exercise If the Air Is Hazy With Wildfire Smoke?" *New York Times*, September 29, 2020, https://www.nytimes.com/2020/09/23/well/move/is-it-safe-to-exercise-if-the-air-is-hazy-with-wildfire-smoke.html.

30. "Wildfire Smoke Has Immediate Harmful Health Effects: UBC Study," University of British Columbia, June 24, 2020, https://news.ubc.ca/2020/06/24/wildfire-smoke-has-immediate-harmful-health-effects-ubc-study/.

31. Mike Hytner and Jonathan Howcroft, "Smoke Plays Havoc as Australian Open Qualifier Suffers Coughing Fit," *Guardian*, January 13, 2020, https://www.theguardian.com/sport/2020/jan/14/safety-first-as-australian-open-qualifying-delayed-due-to-poor-air-quality.

32. George Ramsay, John Sinnott, and Amanda Davies, "Dalila Jakupovic Hits Out at Australian Open Organizers after Bushfire Smoke Forces Her to Quit," CNN, January 15, 2020, https://www.cnn.com/2020/01/14/tennis/australian-open-air-quality-spt-intl/index.html.

33. Emily H., "Wildfire Smoke Forces Players to Withdraw from Australian Open," AmericanUpbeat.com, https://americanupbeat.com/sports/wildfire-smoke-forces -players-to-withdraw-from-australian-open.

34. "Power Output during Exercise," blog post, Body Transform, http://bodytransform .co/Blog/Power+output+during+exercise.html.

35. "Power to Weight Ratio: Watts per Kilogram Explained and How to Boost Yours," *Cycling Weekly*, February 3, 2020, https://www.cyclingweekly.com/fitness/training /the-importance-of-power-to-weight-and-how-to-improve-yours-164589.

36. Carly Stern, "The Fitness Industry Wants to Clean the Air," OZY, September 26, 2019, https://www.ozy.com/the-new-and-the-next/the-next-gym-frontier-air-purifying -workouts/96642/.

37. Adam Boesel, "The World Needs to Learn the Value of a Watt," CleanTechnica, December 17, 2015, https://cleantechnica.com/2015/12/17/the-world-needs-to -learn-the-value-of-a-watt/.

38. "The Future Has Arrived," Green Microgym, https://www.thegreenmicrogym.com /electricity-generating-equipment-2/.

39. "The Story of the Green Microgym," Green Microgym, https://www.thegreen microgym.com/the-story-of-the-green-microgym/.

40. "The World Is Made of Energy. Keep It Flowing!" Green Microgym, https://www .thegreenmicrogym.com/the-story-of-the-upcycle-eco-charger/.

41. Homepage, SportsArt, https://www.gosportsart.com/.

42. Shane McGlaun, "Researchers Create a Wearable Microgrid to Power Gadgets," SlashGear, March 12, 2021, https://www.slashgear.com/researchers-create-a-wearable -microgrid-to-power-gadgets-12663428/.

43. Homepage, UC San Diego Jacobs School of Engineering, https://jacobsschool.ucsd .edu/news/release/www.qualcomm.com.

44. Jennifer Chu, "Footwear's (Carbon) Footprint," MIT News, May 22, 2013, https:// news.mit.edu/2013/footwear-carbon-footprint-0522.

45. Jessica Lyons Hardcastle, "Running Shoes' Carbon Footprint 'More than 60% Manufacturing,'" Environment + Energy Leader, May 23, 2013, https://www.environmentalleader.com/2013/05/running-shoes-carbon-footprint-more-than-60-manufacturing/.

46. "Athletic Footwear Market Worth $95.14 Billion by 2025," Grand View Research, April 2018, https://www.grandviewresearch.com/press-release/global-athletic-footwear-market.

47. Tansy E. Hoskins, "'Some Soles Last 1,000 Years in Landfill': The Truth about the Sneaker Mountain," *Guardian*, March 21, 2020, https://www.theguardian.com/fashion/2020/mar/21/some-soles-last-1000-years-in-landfill-the-truth-about-the-sneaker-mountain.

48. Holly Hummel, "Do Just One Thing: Give Old Shoes New Life," Recycling Advocates, August 23, 2018, http://www.recyclingadvocates.org/do-just-one-thing-give-old-shoes-new-life-aug-2018/.

49. "*Runner's World* Shoe Donation," *Runner's World*, August 28, 2006, https://www.runnersworld.com/gear/a20850552/learn-how-to-donate-used-running-shoes/.

50. "The 2020 Conscious Fashion Report," Lyst Insights, https://www.lyst.com/data/2020-conscious-fashion-report/.

51. Ibid.

52. "How Much Do Our Wardrobes Cost to the Environment?" World Bank, September 23, 2019, https://www.worldbank.org/en/news/feature/2019/09/23/costo-moda-medio-ambiente.

53. Sonu Trivedi, "Sustainable Fashion Movement Explores Dark Side of the Industry," WUSF Public Media, April 17, 2019, https://wusfnews.wusf.usf.edu/environment/2019-04-17/sustainable-fashion-movement-explores-dark-side-of-the-industry.

54. "Too Many Clothes? These 9 Reasons Show Why Your Closets Are Packed," blog post, Organized Interiors, https://www.organizedinteriors.com/blog/too-many-clothes/.

55. Morgan McFall-Johnsen, "These Facts Show How Unsustainable the Fashion Industry Is," World Economic Forum, January 31, 2020, https://www.weforum.org/agenda/2020/01/fashion-industry-carbon-unsustainable-environment-pollution/.

56. "Measuring Fashion: Insights from the Environmental Impact of the Global Apparel and Footwear Industries," Quantis, https://quantis-intl.com/report/measuring-fashion-report/.

57. "A Guide to the Most and Least Sustainable Fabrics," Eco-Stylist, June 4, 2020, https://www.eco-stylist.com/a-guide-to-the-most-and-least-sustainable-fabrics/.

58. Priscilla Greene, "Organic Cotton vs. Recycled Cotton—Sustainable Fashion's Many Faces," TotebagFactory.com, February 17, 2020, https://totebagfactory.com/blogs/news/organic-cotton-vs-recycled-cotton.

59. Ibid.

60. "Hemp: Facts on the Fiber," OrganicClothing.blogs.com, December 7, 2005, https://organicclothing.blogs.com/my_weblog/2005/12/hemp_facts_on_t.html.

61. Caroline Lennon, "5 Eco-Friendly Fabrics to Have and to Hold," One Green Planet, https://www.onegreenplanet.org/lifestyle/5-eco-friendly-fabrics-to-have-and-to-hold/.

62. "Fabrics," Orange Fiber, http://orangefiber.it/en/fabrics/.

63. "Fabrics from Oranges—Interesting and Exciting," Unnati Silks, https://www.unnatisilks.com/blog/fabrics-from-oranges-interesting-and-exciting/.

64. Ibid.

65. Homepage, Orange Fiber, http://orangefiber.it/en/.

66. Erik van Sebille, "How Much Plastic Is There in the Ocean?" World Economic Forum, January 12, 2016, https://www.weforum.org/agenda/2016/01/how-much-plastic-is-there-in-the-ocean/.

67. "Fact Sheet: Single Use Plastics," EarthDay.org, March 29, 2018, https://www.earthday.org/fact-sheet-single-use-plastics/.

68. "Conserve Water," Boston University Sustainability, http://www.bu.edu/sustainability/how-to/conserve-water/.

Conclusion

1. "Racial Disparities and Climate Change," Princeton Student Climate Initiative, August 15, 2020, https://psci.princeton.edu/tips/2020/8/15/racial-disparities-and-climate-change.

ABOUT THE AUTHOR

Bonnie Schneider is a national television meteorologist based in New York City, appearing on MSNBC/NBC News and Yahoo! Finance. She created the platform Weather & Wellness, successfully launching its original video content focusing on climate change and health for New York–based *Newsday*'s digital site. Bonnie connects with her fans and answers their weather questions through Facebook, Twitter, and Instagram. Bonnie has provided on-camera insight and expertise on everything from hurricanes to snowstorms for CNN, HLN, Bloomberg TV, and the Weather Channel.